ADVANCES IN SORGHUM SCIENCE

Botany, Production, and
Crop Improvement

ADVANCES IN SORGHUM SCIENCE

Botany, Production, and Crop Improvement

Ratikanta Maiti, PhD, DSc
Humberto González Rodríguez, PhD
Ch. Aruna Kumari, PhD
Sameena Begum, MSc

First edition published 2022

Apple Academic Press Inc.
1265 Goldenrod Circle, NE,
Palm Bay, FL 32905 USA
4164 Lakeshore Road, Burlington,
ON, L7L 1A4 Canada

CRC Press
6000 Broken Sound Parkway NW,
Suite 300, Boca Raton, FL 33487-2742 USA
2 Park Square, Milton Park,
Abingdon, Oxon, OX14 4RN UK

© 2022 Apple Academic Press, Inc.

Apple Academic Press exclusively co-publishes with CRC Press, an imprint of Taylor & Francis Group, LLC

Library and Archives Canada Cataloguing in Publication

Title: Advances in sorghum science : botany, production, and crop improvement / Ratikanta Maiti, PhD, DSc, Humberto González Rodríguez, PhD, Ch. Aruna Kumari, PhD, Sameena Begum, MSc.

Names: Maiti, R. K., 1938- author. | Maiti, R. K., 1938- author. | González Rodríguez, Humberto, 1959- author. | Aruna Kumari, C. H., 1972- author. | Begum, Sameena, author.

Description: First edition. | Includes bibliographical references and index.

Identifiers: Canadiana (print) 2021011262X | Canadiana (ebook) 20210112662 | ISBN 9781771889674 (hardcover) | ISBN 9781774637449 (softcover) | ISBN 9781003127628 (ebook)

Subjects: LCSH: Sorghum. | LCSH: Crop improvement.

Classification: LCC SB191.S7 M35 2021 | DDC 633.1/74—dc23

Library of Congress Cataloging-in-Publication Data

CIP data on file with US Library of Congress

ISBN: 978-1-77188-967-4 (hbk)
ISBN: 978-1-77463-744-9 (pbk)
ISBN: 978-1-00312-762-8 (ebk)

About the Authors

Ratikanta Maiti, PhD, DSc
*Formerly Visiting Research Scientist, Forest Science Faculty,
Autonomous University of Nuevo Leon, Mexico*

Ratikanta Maiti, PhD, DSc, was a world-renowned botanist and crop physiologist. He worked on jute and allied fibers at the former Jute Agricultural Research Institute (ICAR), India. He also worked as a plant physiologist on sorghum and pearl millet at ICRISAT (International Crops Research Institute for the Semi-Arid Tropics) for 10 years. He was a professor and research scientist at three different universities in Mexico. He also worked for six years as a Research Advisor at Vibha Seeds, Hyderabad, India, and as a Visiting Research Scientist for five years at the Universidad Autonoma de Nuevo Leon, Facultad de Ciencias Forestales, Nuevo Leon, Mexico Mexico. As the author of more than 40 books and about 500 research papers, he won several international awards, including an Ethno-Botanist Award (USA) sponsored by Friends University, Wichita, Kansas, the United Nations Development Programme; a senior research scientist award offered by Consejo Nacional de Ciencia y Tecnología (CONACYT), Mexico, and gold medal from India 2008 offered by ABI. He was Chairman of the Ratikanta Maiti Foundation and chief editor of three international journals. Dr. Maiti died in 2019.

Humberto González Rodríguez, PhD
*Faculty Member, Autonomous University of Nuevo Leon, Facultad de
Ciencias Forestales (School of Forest Sciences), Nuevo Leon, Mexico*

Humberto González Rodríguez, PhD, is a faculty member at the Universidad Autonoma de Nuevo Leon, Facultad de Ciencias Forestales (School of Forest Sciences), Nuevo Leon, Mexico. He received his PhD in Plant Physiology from Texas A&M University in 1993 under the advice of Dr. Wayne R. Jordan and Dr. Malcolm C. Drew. He is currently working on water relations and plant nutrition in native woody trees and shrubs, northeastern Mexico. In addition, his research includes nutrient deposition via litterfall in different forest ecosystems. Dr. Rodríguez teaches chemistry, plant physiology, and statistics.

Ch. Aruna Kumari, PhD

Assistant Professor, Department of Crop Physiology, Agricultural College, Jagtial, Professor Jaya Shankar Telangana State Agricultural University (PJTSAU), India

Ch. Aruna Kumari, PhD, is an Assistant Professor in the Department of Crop Physiology at Agricultural College, Jagtial, Professor Jaya Shankar Telangana State Agricultural University (PJTSAU), India. She has nine and half years of teaching experience at PJTSAU and seven years of research experience at varied Indian Council of Agricultural Research institutes as well as at Vibha Seeds. She is the recipient of a Council of Scientific & Industrial Research fellowship during her doctoral studies and was awarded a Young Scientist Award for Best Thesis Presentation at the National Seminar on Plant Physiology. She teaches courses on plant physiology and environmental science and has taught seed physiology and growth, yield, and modeling courses. She is the author of book chapters in *Advances in Cotton* and *Chili and Food Security for Sustainable Development.* She was one of editors of the book Glossary of Plant Physiology and editor of five books, *Advances in Bio-Resource and Stress Management* and *Applied Biology of Woody Plants.* Experimental Ecophysiology and Biochemistry of Trees and Shrubs, Advances in Cotton Science: Botany, Production, and Crop Improvement and Advances in Rice Science: Botany, Production, and Crop Improvement. She had also published many research publications in national and international journals.

Sameena Begum, MSc

Researcher

Sameena Begum is a young researcher. She completed a BSc degree in Agriculture with distinction and an MSc degree in Genetics and Plant Breeding with distinction from the College of Agriculture, Professor Jayashankar Telangana State Agricultural University, Hyderabad, India. During her master's degree program, she conducted research on the combining ability, gall midge resistance, yield, and quality traits in hybrid rice (*Oryza sativa* L.) and identified two highly resistant hybrids. She has been an author of book chapters in three books, *Experimental Ecophysiology and Biochemistry of Trees and Shrubs*, *Advances in Cotton Science: Botany, Production, and Crop Improvement*, and *Advances in Genetics and Plant Breeding*. She was also one of the editors of the books, *Advances in Cotton Science: Botany, Production, and Crop Improvement* and *Advances in Rice Science: Botany, Production, and Crop Improvement*. She had also published five research publications in national and international journals from her research work.

Contents

Abbreviations

ABA	abscisic acid
ABTS	2,2'-azinobis (3-ethyl-benzothiazoline-6-sulfonic acid)
AFLP	amplified fragment length polymorphism
AFPs	antifungal proteins
AGDW	aboveground dry weight
AMF	arbuscular mycorrhizal fungi
ANOVA	analysis of variance
AQPs	aquaporins
AUSPC	area under striga progress curve
BAC	bacterial artificial chromosome
BMR-6	brown-midrib
BNF	biological nitrogen fixation
BNIs	biological nitrification inhibitors
BUK	Bayero University Kano
CA	concentration addition
CAD	cinnamyl alcohol dehydrogenase
CDA	canonical discriminant analysis
CDD	conserved domain database
CER	carbon dioxide exchange rate
CEY	carbohydrates and ethanol yield
CIM	composite interval mapping
CMS	cytoplasmic male sterility
COMT	caffeic acid O-methyltransferase
CWC	cell wall content
CZ	central zone
DAA	days after anthesis
DAP	days after pollination
DAPC	discriminant analysis of principal components
DDGS	dried distiller's grain with solubles
DM	dry matter
DP	diastatic power
DPPH	2,2-diphenyl-1-picrylhydrazyl
DSC	differential scanning calorimetry

EC	exotic cultivar
EMS	ethyl methanesulfonate
EPG	electrical penetration graph
ESIP	Ethiopian Sorghum Improvement Program
FAN	free amino nitrogen
FNR	false negative rate
FTN	fertile tiller number
FVs	farmer varieties
GA	gibberellic acid
GBS	genotyping-by-sequencing
GC	gas chromatography
GD	genetic diversity
gDNA	genomic-DNA
GE	genetic erosion
GFP	green fluorescence protein
GIS	geographic information system
GLAM	green leaf area at maturity
GMR	grain mold-resistant
GMS	genetic male-sterility
GMS	grain mold-susceptible
GWAS	genome-wide association study
HMT	heat-moisture treatment
HNT	high night temperature
HPLC	high-performance liquid chromatography
HR	hypersensitive response
HRT	hydraulic retention times
HT	high temperature
HTPP	high throughput plant phenotyping
IA	independent action
IC	indigenous cultivar
ICRISAT	International Crops Research Institute for the Semi-Arid Tropics
IR	incompatible response
ITS	internal transcribed spacer
IVPD	*in vitro* protein digestibility
IVs	improved varieties
LAI	leaf area index
LC	liquid chromatography

LG	linkage group
LID	limited irrigation-dry
LMX	late metaxylem
LTF	lead-tolerant fungus
LWP	leaf water potentials
Lys	lysine
MAB	marker-assisted breeding
MAS	marker-assisted selection
MDR	model deviation factor
MDS	multidimensional scaling
MF	methylation filtration
MNR	menthone: neomenthol reductase
MQM	multiple QTL mapping
MSG	monosodium glutamate
MTBE	methyl tert-butyl ether
MWE	mulberry water extract
N	nitrogen
NDVI	normalized difference vegetation index
NIL	near isogenic line
NT	no-tillage
OM	organic matter
ORAC	oxygen radical absorbance capacity
OT	optimum temperature
P	phosphorus
PaO	pheophorbide *a* oxidase
PAR	photosynthetically active radiation
PCR	polymerase chain reaction
PDS	phytoene desaturase
PHS	pre-harvest sprouting
PM	plasma membrane
PPB	participatory plant breeding
PT	paratill
PV	peak viscosity
QTL	quantitative trait loci
RAPD	random amplified polymorphic DNA
RCBD	randomized block design
rDNA	ribosomal DNA
RFLP	restriction fragment length polymorphism

RI	recombinant inbred
RIL	recombinant inbred line
RIP	ribosomal inactivating proteins
RLD	root length density
ROS	reactive oxygen species
RS	resistant starch
RSA	root system architecture
Rst	rate of shear thinning
S	sulfur
SAP	sorghum association panel
SbFT	sorghum flowering locus T
SCARs	sequence characterized amplified regions
SCZ	south-central zones
SD	short-day
sgRNA	single guide RNA
SKCS	single kernel characterization system
SLN	specific leaf nitrogen
SM	stubble mulch-tillage
SNP	single nucleotide polymorphism
SRWC	short-rotation woody crops
SSR	simple sequence repeat
SSU	small subunit
ST	sweep-tillage
SWE	sorghum water extract
SWZ	south-west zone
TAA	total amino acids
TADD	tangential abrasive dehulling device
TEAA	total concentration of essential amino acids
TGA	thermogravimetric analyzer
TGMRs	threshed grain mold ratings
TLN	total leaf number
Tp	temperature
TPLA	total leaf area per plant
TPR	true positive rate
TRIM	tandem repeats in miniature
TT	thermal time
UAS	unmanned aerial systems
UAV	unmanned aerial vehicles

VAMF	vesicular-arbuscular mycorrhizal fungi
VPDs	vapor pressure deficits
WSF	wheat-sorghum-fallow
WUE	water use efficiency
Zn	zinc

Acknowledgments

The authors sincerely thank Apple Academic Press for accepting the manuscript and publishing this book within the prescribed timeline.

We heartily thank our chief author Dr. Ratikanta Maiti, an eminent dedicated scientist, for his continuous efforts, motivation, and initiation in writing this book and making this book unique, and bringing excellent shape for the next generations.

Preface

Sorghum is a vital crop as a source of food and forage and has many other uses in semiarid regions of the world. The present book deals with various aspects of sorghum crop starting from its origin, its domestication and up to biotechnology, written on the basis of world literature up to 2019.

Sorghum is a cultivated tropical cereal grass that was thought to be first domesticated in 1000 BC in North Africa. It has three distinct lineages. It is grown in warmer climatic areas across the world, occupying fifth place in importance among the cereals. Of the world's annual production of ~60 million tons, approximately 20 million tons are produced in Africa. For millions of people residing in at least of the 30 countries of Africa and Asia, it acts as a staple dietary food. It also acts as a source of fodder and as food to small farm-holding sector farmers. It is a rich source of micronutrients (minerals and vitamins) and macronutrients (carbohydrates, proteins, and fat). Its presence of resistant starch (RS) makes it a suitable food for obese, diabetic, and gluten allergic people. Malts prepared from sorghum have tremendous activities of α-amylase and β-amylase, making it useful for various agro-industrial foods. As an excellent source of starch and protein, several industrial products occur as processed products, such as flour, grits, and flakes, malted foods, beverages, and beer.

The authors have provided information on almost all aspects of several types of sorghum under one umbrella, in this volume, *Advances in Sorghum Science: Botany, Production, and Crop Improvement*. This book covers most the aspects of sorghum, starting from the background of sorghum crop, to its production, origin to domestication, ideotype, botany, physiology of crop growth and productivity, abiotic and biotic factors affecting crop productivity, methods of cultivation, postharvest management, grain quality analysis to food processing, improvement of sorghum crop, research advancements in breeding and biotechnology up to 2019. The content of each chapter describes the topic extensively and is enriched with the recent research literature.

This book was written in a lucid style to help undergraduates, graduates, academicians, and teaching faculty to gain knowledge and understand the crop. This book especially guides researchers working on sorghum and sorghum scientists to understand the relationship between several disciplines and the implementation of new methods and technology covered in recent on sorghum crop for crop improvement for higher productivity. A multi-pronged approach needs to be used to increase sorghum productivity to meet the world's demands.

Most of the books published on sorghum are on specific aspects; very few books attempt to bring together all disciplines in a concrete form like the present book.

Some problems were faced during the course of writing this book. However, the authors' dedication and determination played a major driving force in overcoming the problems and successfully completing the book.

CHAPTER 1

Background and Importance of Sorghum

ABSTRACT

Sorghum is a major cereal crop for food and feeds with high nutritional values. It is processed to produce a number of convective and nutritious traditional foods. In this chapter, sorghum's background and importance (including its various industrial products) at the global level are presented and sorghum growing environments and sorghum production potentials are discussed.

1.1 BACKGROUND AND IMPORTANCE OF SORGHUM

Sorghum is a major crop used for food and feed and it has high nutritional values. It is generally adapted to the semiarid climatic sections of the world. It is the second most cereal meeting the food requirements of a large population in Africa. In Africa alone, 20 million tons of sorghum is produced per annum, which is equivalent to nearly one-third of the world's sorghum production. It is well adapted in Africa, to its climatic conditions, drought, and waterlogging conditions. It is processed to make several convective and nutritious traditional foods, such as semi-leavened bread, couscous, dumplings, and fermented and non-fermented porridges and traditional African beer. For many of the African people sorghum serves as a potential source of economy (Talor, 2011).

In the smallholder farming sectors or in traditional farming or under high input commercial farming sectors, sorghum is cultivated both for feed as well as for the production of biofuel. It is a good biofuel crop also. The statistical data analysis has shown that in more than 80% of the global sorghum area, there is the occurrence of low yields. However, it has shown that in the rest of the developed world where it is under cultivation higher yields were realized. It is cultivated in more than 100 countries, and produces more than 60% of world sorghum production. It occupies a vital staple food grain in large parts of the continent. Regardless of its high economic importance, it is found that the growing area of sorghum all throughout the world has shown a drastic decline in the last four decades with a decline rate of over 0.15 million ha per year. However, in some countries like Brazil, Ethiopia, Sudan, Australia, Mexico, Nigeria, and Burkina Faso the sorghum cropped areas have increased. This increase in sorghum-cropped area is due to the bringing up of new lands under sorghum cultivation or part of the areas is diverted towards sorghum cultivation while the rest of the area was utilized for the cultivation of wheat and maize. They mentioned that global sorghum production peaked during the mid-1980s; it is reduced to about 13–15 in almost all the sorghum-growing regions but Africa. Conversely, consumption of sorghum for food purposes reduced due to a change in food habits and consumer preference for economic status, while animal feed and other industrial purposes are increasing (Hariprasanna et al., 2014).

1.1.1 FOOD VALUE

Sorghum, as a vital millet crop, provides good sources of protein. The consumption of sorghum and millets by millions of people at Africa has enabled in the achievement of food security in the region (Peter et al., 2004). It forms a major dietary staple food for millions of people in and across 30 countries of Africa and Asia's subtropical and semi-arid regions. It is both a source of food and fodder (Hariprasanna et al., 2014).

Unlike other cereals as barley, rice, or wheat sorghum can effectively grow in the world's semi-arid regions. It is also an excellent source of protein and starch like other cereal grains. It can also be processed into different products like flour, starch, grits, and flakes and a large variety of industrial products. Further, it can be processed to make the malted foods, beverages, and beer (Palmer, 1992).

1.1.2 FORAGE VALUE

A change in lignin composition of leaf is much helpful in improving the forage digestibility for livestock. Mutations in sorghum loci *BMR 6* and *BMR 12* had produced some sorghum phenotypes possessing brown colored leaf midrib veins. These were associated with reduced lignin content in them. The two loci in sorghum encodes the enzymes cinnamyl alcohol dehydrogenase (CAD) and a caffeic O-methyl transferase (COMT). These two enzymes are important in the monolignol biosynthesis. The hybrids that were generated from these lines had ameliorated yields and the forage had increased digestibility due to the maintenance of required lignin content, though some isogenic lines of sorghum exhibited yield reductions (Scott et al., 2010). An exceptional defect in sorghum in wet cooking is a decline in the protein digestibility (Taylor, 2011).

1.1.3 SOURCE OF BIOENERGY

Sorghum has the capacity of production of lignocellulose, starch, and sugar. In the present day conditions there is an increase in cost of energy and oil and gas resources are being replenished at a quick pace. There is a need for the development of some alternate fuel sources from the renewable sources, as the existing oil and gas reserves are the finite resources. Ethanol has been developed as a renewable transportation fuel. Its production at present is either sugar or starch-based. Owing to the fact that these carbohydrates have their usage as sources of food and feed, there is a requisite to the generation of a large and sustainable supply of the biomass that would enable in the preparation of biofuels. Biofuel production from crops having lignocellulose would be more cost-effective which can be used for bioenergy production. Sorghum appears to be one such efficient crop with lignocelluloses content in the production of profitable ethanol as a biofuel (William et al., 2007).

In Africa, sorghum is grown mainly for meeting the food requirements of the people. Some of the important biochemical constituents are a starch (amylose and amylopectin). Sorghum also has higher activities of starch depolymerizing enzymes. Owing to its greater biodiversity, it has ability in meeting the agricultural and food requirements of a vast population and effectively can ensure the food security. Not only it is rich in starch, it is also a rich source of macronutrients (carbohydrates, proteins, and fat) and micronutrients (minerals and vitamins). Sorghum is good for consumption by diabetic people and also the obese persons due to its resistant starch (RS) content. People who were allergic to gluten content of wheat can have alternate source of food from sorghum. In Africa, sorghum is given much emphasis for production because of its wider utility of starch and its degrading enzymes that are helpful in the preparation of sorghum foods in Africa, e.g., "To Thin porridges for infants," granulated foods "*couscous*" local beer "dolo," as well agro-industrial foods such as lager beer and bread (Mamoudou et al., 2006).

Most of the whole grain products are proposed as healthy diets, because they are rich sources of dietary fiber and antioxidant substances. Among the four cereals whole grain products of barley, pearl millet, rye, and sorghum which were analyzed for composition of dietary fiber, RS, minerals, and total phenols and antioxidant properties; it was observed that the adapted grains had better nutritional quality than the commercially available hard and soft wheat flours or their products. All these products were rich in RS, soluble, and insoluble dietary fibers, minerals, and antioxidants. However, barley had the highest levels of phosphorus, calcium, potassium, magnesium, sodium, copper, and zinc, the second-highest content of iron, and these were found in millet. Sorghum had exceptionally high antioxidant activities. The nutritional data has suggested that barley and sorghum, holds as healthy food ingredient (Sanaa et al., 2006). It can be cultivated efficiently in the semi-arid regions of the world accompanied by other cereal crops.

Further, in some world's poorest countries, it can contribute significantly to the agricultural development if improvements are made in the relevant science to produce better quality sorghum and improved yields. Advances in the knowledge among the locally-based industries for production of quality based varied industrial products unlike those prepared from cereals like barley, wheat, and rice, the products of sorghum can also be economical in the semi-arid regions of the world. However, for a long time it has been used traditionally for producing foods, malt beverages, and beer. Its structure and function have not been studied to the same level as grains that grow readily in the world's more developed regions (Sanaa et al., 2006).

In addition to its richness in starch, macro, and micronutrient sorghum, it also appears a novel food source serving the requirements of neutraceuticals such as antioxidant phenolics and cholesterol reduction waxes. Snack foods, cakes, cookies, pasta, parboiled rice-like products are also prepared from sorghum. The production of wheat-free sorghum or millet bread appears to be a key challenge. However, this could be easily prepared with the addition of some additives like native, pregelatinized starches, hydrocolloids, egg, fat, and rye pentosans for increasing the bread quality. Commercial beers like Lager and Stout beers are prepared from sorghum. Despite its utility, sorghum malt is inefficient in replacing the barley malt completely owing to its high starch gelatinization temperature and low *beta*-amylase activity. The research on utilization of sorghum for bioethanol and other bio-industrial products has currently been concentrated only on the improvement of the economics of the bioethanol extraction process through cultivar selections. Research has to be concentrated on the development of methods for the extraction of bioethanol from low-quality grain and recovery of some valuable products from the pre-processed sorghum products during the bioethanol production. Some of these by-products like kafirin prolamin proteins and pericarp waxes have the potential to be used as in preparation of bioplastic films and as coatings for foods, mainly because of their hydrophobicity (Taylor et al., 2006).

Gluten protein is found in wheat and other closely associated cereals as barley and rye. Its consumption causes Celiac disease, where in there is an autoimmune reaction to gluten proteins, a common condition seed in the genetically predisposed people suffering from Celiac disease. There is destruction of the villi in the small intestine. The destruction of villi in the small intestine induces the malabsorption of nutrients. It further induces other gluten induced autoimmune diseases. Sorghum flour made from white sorghum hybrids is light in color and its neutral taste that does not impart odd colors or flavors to food products. It is desirable for its use in wheat-free food products. Consumption of sorghum food is thus considered as a safe food for those patients suffering from Celiac disease. Though sorghum flour is in some way related to maize flour, its gluten intolerance ability was not tested directly. Thus, there is a need to evaluate its safety and tolerability in celiac patients (Carolina et al., 2007). There is a need to modify food of kafirins and develop nutritious, palatable, and inexpensive staple food like bread and pasta to suit celiac patients' food requirements. Grain sorghum and its proteins are safe for celiac patients and individuals with variable levels of gluten intolerances (Normell et al., 2010).

1.2   SORGHUM ETYMOLOGY

Molecular and morphological analysis of off-type sorghum taxa has shown that three distinct lineages are evident among these off types in sorghum. Though the relationships among the lineages were not resolved, each lineage has represented a distinct genus with wide differences in the Andropogoneae tribe. The type species for the name *Sorghum* is *S. bicolor*, the cultivar. *S. halepense* and *S. nitidum* are also kept in *Sorghum*. The name *Sarga* is resurrected and it encompasses a group of species that previously made up the bulk of subgenera *Parasorghum* and *Stiposorghum*. A new genus, *Vacoparis*, includes cytologically and morphologically diverse Australasian taxa, *V. macrospermum* and *V. laxiflorum*. The taxonomy put forward demonstrated a less number of name changes in the rankless alternative. These changes are associated where consideration of constraints of rank is meager. Among the three lineages there are uncertain relationships. Further, a vast number of taxa are found to compose the members of subtribe Saccharinae. This has demonstrated the difficulties that existed to designate rank to taxa. New data may bring about future name changes dramatically. Rankless classification has progressed and validated and its utility was specified with several concrete examples. Rankless classification enables in provision of an insight into those specific cases that enable in the evaluation of the weakness and strengths. The rankless classification has shown that the species boundaries in *Sarga* are quite different to those which were previously defined. In the past classifications the taxa were differentiated based on the characters that exhibited a continuous variation among the specimens. This has led to a constraint in expansion of the species. Therefore, only a few species which were morphologically variable were only known (Russel, 2003).

The incidence and rate of spontaneous crop-to-weed hybridization between *Sorghum bicolor* and *S. halepense*, Johnson grass was examined. In progeny testing an isozyme marker was used to identify hybrid plants. Significant variations were observed in the incidence and rate of hybridization with respect to the weed distance from the crop, location of the study site, and year the study. Crop/weed hybrids were found at a distance of 0 5–100 m from the crop. This indicated that the interspecific hybridization can occur in this system at a substantial and measurable rate and transgenes introduced into crop sorghum might have the opportunity to escape cultivation through interspecific hybridization with Johnson grass (Paul et al., 1996).

1.3 SORGHUM GROWING ENVIRONMENT

1.3.1 SORGHUM AND ENVIRONMENT INTERACTION

Environments have key role on growth and development of sorghum. Physiological research aiming at crop improvement mostly concentrates in understanding the interactions that occur between temperature and radiation use efficiency of sorghum crop. This helps in understanding the interaction effect of environment on genotype. It enables in improvement of the efficiency of the models in interfacing the physiological research for crop improvement. The photosynthetically active radiation (PAR) intercepted by genotype is used for assessing genotype-by-temperature interaction that exists. The total dry matter (DM) produced by a genotype is generally considered as a product of intercepted PAR and radiation use efficiency. The aboveground DM produced from a genotype is estimated from total DM and partitioning between tops and roots. Sorghum [*Sorghum bicolor* (L.) Moench] hybrids RS610 and ATx623/RTx430 grown at 17°C and 25°C in glasshouses and were monitored for DM by sequential harvesting, and radiation use efficiencies and temperature interactive studies were done by Hammer and Vanderlip (1989). The DM production of RS610 and ATX623/RTx430 did not show any difference at either temperature, although there was high DM production at 26°C by ATx623/RTx430. The partition of DM to roots was high in ATx623/RTx430 (0.25) than RS610 (0.22). This effect on partitioning emphasized the potential error in comparing radiation use efficiencies among genotypes by means of computations based only on aboveground DM.

Variations in temperature and rainfall account for more than half of the environment and genotype X environment interactions. Three growth stages of sorghum viz., planting to panicle initiation (GS1), panicle initiation to anthesis (GS2), and anthesis to physiological maturity (GS3), were estimated with three periods of crop seasons; each with an equal number of growing degree-days in grain sorghum has shown that the above variables added to differential genotypic response to environments. These responses differed within the maturity groups and growth stages. It was observed that temperature was the key factor that affected the environmental variability for yield and also the seed number. The effects of temperature and rainfall in GS2 and GS3 were highly correlated with GE interaction effects for yield in all maturity groups. Minimum temperature was found to be of much importance rather than the maximum temperature, particularly for the early and the late-maturing genotypes. There was more contribution to rainfall by the preseason precipitation which had its contribution to the GE interaction sums of square for yield and seed number. This interaction effect was opposite for seed weight. In late maturing genotypes in comparison to the early and medium maturing genotypes, there was more contribution to GE interactions for yield by the variations in the weather factors (Mohammad and Francis, 1984).

Genotype and environment interaction is also seen at the individual leaf area dynamics and its effect on grain yield. Leaf area of tillering sorghums was anticipated on the basis of the appearance and expansion of individual leaves. Modeling of genotypic and environmental control of leaf area dynamics in grain sorghum at the individual leaf level through the explanation of a framework that simulated the total leaf area of tillering sorghums, on the prediction of the emergence and expansion of individual leaves on grain sorghum hybrids cultivated at regions in latitude from 39° 11′N to 27° 33′S. A leaf area model was established that simulated axes within plants separately. The production and expansion of individual leaves were simulated on the main culm and each emerging tiller. The model simulated the total leaf area per plant (TPLA) (without losses due to senescence). The model through its functions enabled in the prediction of development of successive axes on plants as a function of thermal time (TT) from emergence; a constant rate of mature leaf production per axis per unit of TT; the profile of mature leaf areas for leaves on each axis a function of total number of leaves produced per axis; and the leaf area contribution of immature leaves at any time as equivalent to that for a constant number of mature leaves. The general model simulated TPLA over time, which accounted for 90% of their observed variation with a root mean square deviation of 905 cm^2 for observed values of TPLA varying from 161 cm^2 to 10,584 cm^2 for all the seven sorghum hybrids that were cultivated at three locations (Carberry et al., 1993).

Crop simulation models of plant processes can effectively capture the biological interactions between the sensing of signals at an organ level (e.g., drought influencing roots) and the plant's response at a biochemical level (e.g., variation in development rate). These interactions exhibit their results at the organ (or crop) level, for example, reduced growth. The simulation models perform many roles and enable in studying the complex control of phenotypes like yield. Thus, these can make an index of the climatic environment (e.g., of drought stress) for breeding program trials. In wheat and sorghum grown regions of northern Australia, mid-season drought produces a large genotype by environment interactions. Using gene action to determine the value of input trait parameters to crop models, simulated multi-environment trials computed the yield of 'synthetic' sorghum cultivars cultivated in historical or artificial climates with present or possible management regimes. The biological interactions among traits limit the crop yields to only those that are biologically probable in the assumed set of environments. This permits the formation of datasets that are more 'realistic' representations of genes by trait by environment interactions. The modeling approach has an added advantage in that 'biological and experimental noise' can be manipulated independently. These 'testbeds' for statistical techniques can be extended to the interpretation of a crossing and selection program where the processes of chromosomal recombination are simulated with a quantitative genetics model and applied to the trait parameters (Scott and Chapman, 2008).

Sorghums grown in the subtropics are prone to climate risks as they are likely to be exposed to high rainfall variability in the future years. Quantification of production risks can help farmer decision-making by mechanistic crop simulation models that are simple and enable the assessment of climate risks in the water-limited environments. The mechanistic simulation models can simulate grain yield, accumulation of biomass, crop leaf area, phenology, and soil water balance through the utilization of a daily time-step and readily available weather and soil information with no nutrient limitations. The Crop simulation model of Hammer and Muchow (1994) predicted satisfactorily, for 94% and 64% of the variation in total biomass and grain yield, respectively.

Temperature and photoperiod also influence the ontogeny of sorghum. Novel genotypes are thought to be better adapters to tropical environments. Photoperiods greater than 13 h led to lengthening of the duration of emergence to floral initiation (GS1) by five days in sorghum genotypes. Sorghum genotypes also exhibited variability in the duration of GS1 by up to 10 d at both temperatures. Further, it was observed that there was a quick earlier response in the sorghum hybrids rather than their parents. This indicates earliness to show some form of dominance. The photoperiod had no effect on the duration of floral initiation to anthesis (GS2), though there was 3 d variability in the hybrids of sorghum grown at sites in Australia and the USA, latitudes from 16 to 39°C. The durations of GS1 and GS2 varied from 17 to 128 d and 24- to 85 d. Development rates of all hybrids showed a curvilinear response to temperature in both phases. There were lower rates of development at all the temperatures in the new hybrids. However, differences seem to be more at higher temperatures (>25°C). All hybrids tested exhibited identical short-day (SD) photoperiodic response in GS1, with a critical photoperiod of 13.2 h (Hammer et al., 1989).

Leaf area dynamics are under the control of influences of genotype and environment. General models for leaf area dynamics of uniculm and tillering sorghum at the whole plant as well as individual leaf levels were developed. Carberry et al. (1993) examined and quantified the genotypic and environmental controls by Green leaf area model that independently assessed the leaf area production and senescence. Accumulation of total plant leaf area (TPLA) (without losses due to senescence) was simulated with a logistic function of TT from emergence and it increased to its maximum value (TPLA$_{max}$) shortly before flowering. To calculate TT, base optimum and maximum temperatures of 11, 30 and 42°C respectively were acquired by assessing the effect of temperature on rate of development of fully expanded leaves. Therefore, TT incorporated the major effect applied on TPLA by temperature. Most residual genotypic and environmental variation in TPLA was associated to variation in TPLA$_{max}$. Values of TPLA$_{max}$ were calculated from total leaf number (TLN) on the main culm and fertile tiller number per plant (FTN) by allowing for a curvilinear increase in TPLA$_{max}$ with TLN and a consecutive reduction in total leaf area produced by following surviving tillers compared with that on the main culm. The potential genotypic and environmental controls on TPLA, introduced via factors affecting TLN and FTN, were taken into account. For seven hybrids grown in eight environments (locations and times of planting), this simple general model accounted for 93% of observed variation in TPLA with time, with a root mean square deviation of 664 cm^2 for noted values of TPLA varying from 161 to 11 302.

1.3.2   WATER DEFICIT EFFECT ON SORGHUM YIELD

Water deficit reduces crop production. It is a general recommendation to grow some drought crops particularly under water deficit conditions to compensate the yield losses. In Northeast Spain sprinkler line source irrigation technique produced a continuous and declining gradient of water.

Under this technique, it was found that in both maize and sorghum the phenology, crop water uptake, total aboveground biomass, and yield were affected. Under well-irrigated conditions there was a greater level of yield decrease by maize genotypes rather than sorghum genotypes. Sorghum yield was greater than maize under moderate or severe water deficits. Sorghum exhibited a higher capacity of extraction of water from deeper soil layers. Its higher yield under irrigation deficit was obtained due to greater aboveground biomass, higher harvest index, and higher water use efficiency (WUE). The findings revealed that sorghum could be a good substitute to maize under water stress conditions in the semi-arid conditions of Northeast Spain (Farre and Faci, 2006).

Climate change has its impact more on the target production systems. Possible adaptation strategies for formulation of the research and developmental policies to meet the challenges of climate change are to be adopted. An analysis of the impact of climate change on the sorghum production system in India with InfoCrop-sorghum simulation model has shown that the influence of climate change would be more on winter crop in central (CZ) and south-central zones (SCZ). In contrast, in the south-west zone (SWZ) the effects appeared to be higher on monsoon crop. Climate change is expected to lessen the grain yields of monsoon sorghum to the rate of 14% in CZ and SWZ and by 2% in SCZ by 2020. The yields appeared to be more affected in 2050 and 2080. Climate change effects on winter crops are projected to decrease yields up to 7% by 2020, up to 11% by 2050 and up to 32% by 2080. Impacts are likely to be more in SWZ region than in SCZ and CZ. On the other hand, the yield loss due to increase in temperature appears to be counterbalanced by a probable increase in rainfall.

Nevertheless, it was suggested that complete amelioration of yield loss beyond 2°C rise may not be achieved even after doubling rainfall in the south-central zone (SCZ) and in the central zone (CZ). Adaptation approaches like such as changing variety and sowing date can decrease the vulnerability of monsoon sorghum to about 10%, 2%, and 3% in CZ, SCZ, and SWZ regions in 2020 scenario. Adaptation approaches decreased the climate change effects and susceptibility of winter crop to 1–2% in 2020, 3–8% in 2050, and 4–9% in 2080. This specifies that more cost-effective adaptation approaches should be carried out to further reduce the net vulnerability of India's sorghum production system (Aditi et al., 2010).

Ramirez-Villegas et al. (2013) mentioned that climate has been changing in the last three decades and will continue to change irrespective of any mitigation strategy. Agriculture is dependent more on climate. Therefore, Agriculture is highly sensitive to climatic fluctuations. The authors gathered the present expert knowledge reported in the FAO-EcoCrop database, using the basic mechanistic model (also named EcoCrop); they further developed the model, giving calibration and evaluation procedures and investigated the impacts of the SRES-A1B 2030s climate on sorghum climatic suitability. The model functioned well, with a high true positive rate (TPR) and a low false-negative rate (FNR) under current conditions when evaluated against national and subnational agricultural statistics (min TPR = 0.967, max FNR = 0.026). The model projected high sorghum climatic suitability in areas where it grows maximum and matched the sorghum geographic distribution comparatively well. Negative effects were projected by 2030s. Vulnerabilities in countries where sorghum cultivation is marginal at present are likely (with a high degree of certainty): the western Sahel region, southern Africa, northern India, and the western coast of India are mainly vulnerable. They highlighted the prospect of using EcoCrop to assess global food security issues, broad climatic constraints and regional crop-suitability shifts in the context of climate change and the likelihood of combining it with other large-area approaches.

The West African semi-arid and sub-humid regions are characterized by high inter-annual rainfall variability. Variable onset of the rainy season occurs at any time during the growing season of sorghum. Climate change is expected to increase this variability of the rainfall. The development of various types with high degrees of heterozygosity and genetic heterogeneity for adaptation traits could help achieve improved individual and population buffering capacity. Traits with potential to improve the adaptive phenotypic plasticity or yield stability in variable climates are photoperiod sensitive flowering, plastic tillering, flooding tolerance, seedling heat tolerance, and phosphorus efficiency. Farmer participatory dynamic gene pool management using broad-based populations and varied selection environments is valuable for developing new diverse germplasm adapted to certain production constraints involving climate variability. Improved cultivars should respond to farmer adoptable soil fertility management and water harvesting techniques to increase sustainable productivity. Large-scale, on-farm participatory testing enables in assessing the varietal performance under evolving climatic variability. It also provides viewpoint on needs and opportunities and increase adoption. Strengthening seed systems will be needed to attain sustainable impacts (Haussmann et al., 2012).

Stay-green, is a drought adaptation mechanism that has been largely accepted for its capacity to improve grain yield and lodging resistance in sorghum [*Sorghum bicolor* (L.) Moench]. Many physiological and genetic studies have been undertaken to work out the relationship of stay green trait and grain yields. Breeding trials data comprising of 1668 unique hybrid combinations from 23 environments producing average yields ranging from 2.3 to 10.5 t ha^{-1} were evaluated for relationship between stay-green trait and yield. The strength and direction of the correlation between stay-green and grain yield exhibited variation in both environment and genetic background (male tester). Most of the correlations were found to be positive. The positive correlations were found especially in environments with yields below 6 t ha^{-1}. However, with an increase in yield trials above 6 t ha^{-1} a trend towards an increased number of negative associations was observed, even though there were higher positive associations with the number and the magnitude of these. Sorghum crop across the world is subjected to post-flowering drought, which lead to production of mean yields of 1.2 and 2.5 t ha^{-1} for the world and Australia. Selection for stay-green in elite sorghum hybrids could be more useful for improving yield under diverse environments (Jordan et al., 2012).

1.3.3 HEAT STRESS ON SORGHUM

In the semiarid regions, Sorghum is very often affected by short periods of high-temperature (HT) stress coinciding during its reproductive development. High temperature (HT) stress is one of the major abiotic stresses of sorghum affecting the sorghum yields. In hybrid of sorghum DK-28E, continuous HT stress led to a delay in the emergence of the panicle apart from its effect on declining the plant height, seed number, seed yield, and harvest indices. HT stress was not to show its effect to a large amount on the leaf photosynthetic rates. A period of short exposure of these hybrid plants for 10 days to heat stress at flowering and also 10 d prior to flowering has resulted in a maximum reductions in seed set and seed yield. However, the heat stress during post-flowering stages (10, 20, and 30 DAF) though reduced the seed yield, a large extent of reduction was seen during the early stages of seed development (Prasad et al., 2008).

HT stress also induces male sterility in grain sorghums. It is a serious problem. In grain sorghums HT has its effect on levels of sugar, the expression profiles of genes associated with sugar-to-starch metabolism in microspore populations. These populations were denoted by pre- and post-meiotic "early" stages through post-mitotic "late" stages that exhibit detectable levels of starch deposition. Starch-deficiency was seen in microspores from HT stress conditions.

Further, they exhibited reduced germination. This accounted for a loss of 27% in seed set. Variations were also seen with respect to sugar profiles, more contrasting variability's were seen with respect to the hexose levels at both "early" and "late" stages at the two temperature regimes. Particularly, in "late" microspores from heat-stressed plants, the sucrose was undetectable and had ~50% lower starch content. Northern blot, quantitative PCR, and immunolocalization data unveiled a significant decline in the steady-state transcript abundance of *SbIncw1* gene and CWI proteins in both sporophytic as well as microgametophytic tissues under HT conditions. In heat-stressed plants, northern blot analyses also showed greatly altered temporal expression profiles of different genes involved in sugar cleavage and utilization (*SbIncw1*, *SbIvr2*, *Sh1*, and *Sus1*), transport (*Mha1* and *MST1*) and starch biosynthesis (*Bt2, SU1, GBSS1*, and *UGPase*). Impairment of CWI-mediated sucrose hydrolysis and successive lack of sucrose biosynthesis may be the most upstream molecular dysfunctions leading to altered carbohydrate metabolism and starch deficiency under high growth temperature conditions in grain sorghums (Mukesh et al., 2007). Mukesh et al. (2010) found that short-term HT stress during the vegetative to reproductive phase transition, for stress-induced male sterility is mainly due to its effect on cell wall invertase that is involved in mediation of sucrose catalysis and microspore meiosis.

Pre and post-meiotic stages were found to be heat sensitive. There was a reduction in the transcriptional activity during the meiotic stage. However, this could be reversed with the provision of optimal growth conditions. In crop plants, male fertility is expressed as a function of pollen production and viability. Plants exhibit variations in pollen production. It is easier to evaluate the pollen production than the quantification of pollen viability. Tuinstra and Jwedel (2000) developed *in vitro* pollen germination assay for sorghum. Among the different germination media substrates, large variations in germination seen in response to varying concentrations of sucrose, boric acid, and calcium nitrate in agar-based media. However, it was found that incubation on agar which has been supplemented with 0.9 M sucrose, 2.43 mM boric acid, and 2.12 mM calcium nitrate was effective for pollen viability. The incubation temperature also had an important role. Variations in temperature between 20 and 40°C had no effect on pollen germination, and at 10°C, the germination was reduced drastically. Reductions in pollen longevity and pollen germination percentage were observed by growth at higher temperature (Prasad et al., 2011).

HTs not only have their effect on pollen viability, HTs that prevail during nighttime also have their influence on the leaf photosynthetic rates and pollen function of grain sorghums. These temperatures might bring about a decrease in the leaf photosynthesis. This is a severe problem in many regions of the arid and semiarid areas on the world. Many of the research statistical data of the climate model projections have shown that in comparison to an increase in the day temperatures, in the forth-coming years there is a possibility of an increase in the night temperatures. Vara and Maduraimuthu (2011) hypothesized that high night temperature (HNT) during flowering would bring about oxidative damage in leaves and pollen grains, resulting in reduced photosynthesis and seed-set. Exposure of the Sorghum plants (hybrid DK-28E) to optimum night temperature (ONT; 32:22°C, day maximum: night minimum) or HNT (32:28°C, day maximum: night minimum) for 10 days after complete panicle emergence, has shown that exposure to HNT resulted in an increase in the membrane damage of thylakoids and there was non-photochemical quenching.

Additionally, it was observed that there was a reduction in the chlorophyll content, quantum yield of PSII, photochemical quenching, electron transport rate and photosynthesis of leaves by HNTs than the optimum night temperature. Likewise, these HNTs have also resulted in increased levels of reactive oxygen species (ROS) level of leaves and pollen grains. An analysis of the lipid molecular species of pollen grains has shown that there was a decline in the saturation levels of phospholipids in the membranes by HNTs. These variations in phospholipids and greater ROS in pollen grains may be accountable for reduced pollen function, leading to lower seed set.

1.4   CULTURAL ROLE OF SORGHUM

In Burkina Faso, there is temporal uncertainty of sorghum yields. This uncertainty also exists with reference to the issues involved in taking key economic decisions. The nature of an economic environment is a very important parameter that has its drastic influence on sorghum production technologies, which in turn influence the valuation of the information and the choice functions. An empirical application to yield analysis has clearly shown this in sorghum grown in Burkina Faso (Jean et al., 1991).

Archaeobotanical studies in the Middle Nile region have conventionally paid attention to domestication processes and the economic/nutritional use of crops. Ethnographic and historical studies have suggested that common practice of converting grain into beer had a great impact on many areas of social life from an early date. However, with in Kushite society, sorghum, and its products, particularly beers, had developed substantial importance in mortuary and other ritual contexts as well as in socioeconomic relations more, by, and large by the late first millennium BC (Edwards, 1996).

Paul et al. (1999) documented the yield increases that occurred in sorghum and also determined the factors that were principally had a main part in the yield increases of sorghum by taking into consideration the annual precipitation, growing-season rainfall, soil water content at planting, soil water use, growing-season evapotranspiration, and year of record they have given the report on assemblage of 502 treatment-years of grain yield data from 37 studies from 1939 to 1997. For the period of 1939–1997 there was an increase in grain yields by about 50 kg ha^{-1} every year. However, during 1956–1997, there was an increase in yields by 139%, with 46 of that percentage unit's resultant from use of improved hybrids, based on results of a consistently managed 40-year study. The remaining 93% units for that period were accredited to other factors, mainly to soil water at planting. Increases in soil water at planting resulted from variations in management practices with time, primarily the adoption of improved crop residue management practices after about 1970.

Grain sorghum is a main rain-fed, Australian summer crop. Earlier works in the region have demonstrated that the sorghum yields in the regions are mostly affected by water limitations. During the growing season, rainfall is associated to the Southern Oscillation, a weather phenomenon that has a strong "memory" and thus the potential for long-range prediction. Differences of sorghum yield have been correlated with values of an index of the

Southern oscillation from during and before the growing season. Southern oscillation index facilitated the early prediction of sorghum yield. Trends in the Southern oscillation index had strong correlations with sorghum yield in the months leading up to the planting season. Estimates made from the trend in the index up to a month before the commencement of the planting season accounted for about 50% of the variance in yield, after removal of long-term yield trends resulting from variations in production technology. The strength of the association, and the early accessibility of the forecasts, seems to offer a potentially useful long-range crop-forecast system (Nicholls, 1985).

Ola et al. (2014) presented molecular evidences of close associations between sorghum population structure and the distribution of ethnolinguistic groups in Africa. They have shown that traditional seed-management practices have played key roles for survival and expansion of agro-pastoral groups in the past, which were still remarkably resilient to threats to human security.

1.5 WORLDWIDE SORGHUM CONSUMPTION

The action of microorganisms or enzymes which bring about desired biochemical variations and major modification to the food is involved in fermented foods. Their production and consumption date back many thousands of years, with first confirmation of barley's alcoholic fermentations to beer and grapes to wine. Food fermentation signifies one of the oldest known applications of biotechnology. This conventional biotechnology has developed from 'natural' processes in which nutrient availability and environmental conditions selected specific microorganisms, through the use of starter cultures, strain improvement, and most recently, gene technology. Lactic acid bacteria, molds of *Aspergillus* spp., *Penicillium* spp. and of Mucorales, and yeasts, often of *Saccharomyces* spp. are most important. Fermented foods are found in diets throughout the world, with dairy, beverage, and cereal products dominating. An important proportion of all diets are represented by fermented foods and beverages worldwide, typically about one-third of food intake, offering a major contribution nutritionally and to flavor and interest in our food consumption.

Country-wise sorghum consumption in the year 2020 is given below.

Rank	Country	Domestic Consumption (1000 MT)
1	China	8,500
2	Nigeria	6,850
3	Ethiopia	5,300
4	Sudan	4,850
5	Mexico	4,700
6	India	4,350
7	United States	3,429
8	Argentina	2,300
9	Brazil	2,200
10	Niger	2,000
11	Burkina Faso	1,800
12	Mali	1,500
13	Cameroon	1,225
14	EU-27	1,022
15	Chad	1,000
16	Bolivia	880
17	Australia	825
18	Tanzania	800
19	Egypt	750
20	Japan	450
21	Uganda	385
22	Togo	300
23	Ghana	280
24	Mozambique	275
25	Kenya	270
26	Senegal	200
27	Saudi Arabia	180
28	Yemen	175
29	South Africa	175
30	Rwanda	170
31	Pakistan	170
32	Eritrea	170
33	Somalia	150
34	Benin	133
35	Uruguay	120
36	El Salvador	100
37	Nicaragua	92
38	Paraguay	75
39	Ukraine	75
40	Haiti	75

FIGURE 1.1   *(Continued)*

41	Zimbabwe	70
42	Mauritania	70
43	Côte D'ivoire	65
44	Guinea	60
45	Taiwan, Province Of China	52
46	Sierra Leone	50
47	Chile	50
48	Thailand	48
49	Guatemala	45
50	Honduras	40
51	Colombia	35
52	Botswana	35
53	Central African Republic	30
54	Gambia	30
55	Iraq	27
56	Burundi	25
57	Philippines	25
58	Zambia	20
59	Iran	20
60	Israel	20
61	Guinea-Bissau	20
62	Venezuela	15
63	Peru	11
64	Ecuador	10
65	Lesotho	10
66	Congo	7
67	Morocco	5
68	Korea	5
69	Swaziland	4
70	Dominican Republic	1

FIGURE 1.1 Sorghum domestic consumption worldwide in 2020 by country (in 1000 metric tons).
Source: United States Department of Agriculture (2020).

Sorghum has ability to produce lignocellulose, sugar, and starch. In view of its history and the present genetic improvement infrastructure available for the species, it is logical to assume that sorghum hybrids for dedicated bioenergy production can be developed in the near-term future and will be cultivated and utilized for bioenergy production (Geofferey, 1994).

The protein quality and digestibility of two high lysine (2.9–3.0 g/100 g protein) and two conventional varieties (lysine content 2.1–2.2 g/100 g protein) of whole grain sorghum milled as flour that were measured through balance studies in 13 children, 6–30 months of age has shown that sorghum protein provided 6.4 or 8.0% of dietary energy. Control diets had 6.4% kcal protein as casein. Children consumed 100–150 kcal/kg body weight/day. Sorghum consumption was correlated with weight loss or poor weight gain. Apparent nitrogen absorption or retention did not show any variation by variety. The average absorption and retention of nitrogen (±SD) from 26-day sorghum dietary periods were $46 \pm 17\%$ and $14 \pm 10\%$ of intake, respectively. Stool weight and energy losses during sorghum periods averaged 2.5 to 3 times control values. After 16 days of sorghum consumption plasma amino acids were ascertained in eleven children. Fasting concentration of total amino acids (TAA) was comparable to values earlier obtained with wheat protein at similar levels of intake. Concentrations of lysine (Lys) and threonine (Thr) were low in total concentration of essential amino acids (TEAA). Analysis of postprandial variations of the Lys/TEAA and Thr/TEAA molar ratios established that Lys was the first limiting amino acid (William et al., 1981).

Bioenergy consumption is highest in countries with high subsidies or tax incentives, such as China, Brazil, and Sweden. The most common forms of bioenergy are conversion of forest residues and agricultural residues to charcoal, district heat, and home heating. Forest residues (including black liquor), bagasse, and other agricultural residues are predominant biomass electric generation feedstocks. Biofuel feedstocks comprise sugar from sugarcane (in Brazil), starch from maize grain (in the US), and oil seeds (soy or rapeseed) for biodiesel (in the US, EU, and Brazil). Total biomass energy consumptions of the six large land areas of the world reviewed (China, EU, US, Brazil, Canada, Australia), amounts to 17.1 EJ. Short-rotation woody crops (SRWC) established in Brazil, New Zealand, and Australia over the past 25 years equal about 50,000 km². In, China SRWC plantings may be in the range of 70,000–100,000 km², whereas in the US and EUSRWC and other energy crops amount to less than 1000 km². With some exceptions (particularly in Sweden and Brazil), the SRWC have been established for uses other than as dedicated bioenergy feedstocks, though, portions of the crops are (or are planned to be) used for bioenergy production. Novel renewable energy incentives, greenhouse gas emission targets, synergism with industrial waste management projects, and oil prices beyond 60 \$ Bbl^{-1} (in 2005) are major drivers for SRWC or energy crop-based bioenergy project (Lynn, 2006).

The production of fermentative hydrogen from the sugars of sweet sorghum extract was examined at different hydraulic retention times (HRT). The consecutive methane production from the effluent of the hydrogenogenic process and the methane potential of the leftover solids after the extraction process were also measured. At the HRT of 6 h highest hydrogen production rate (2550 ml H$_2$/d) was achieved. However, the maximum yield of hydrogen produced per kg of sorghum biomass was attained at the HRT of 12 h (10.4 l H$_2$/kg sweet sorghum). It has been showed that the effluent from the hydrogenogenic reactor is an ideal substrate for methane production with around 29 l CH$_4$/kg of sweet sorghum. After the extraction process, anaerobic digestion of the solid residues yielded 78 l CH$_4$/kg of sweet sorghum. The study established that biohydrogen production can be very effectively combined with a subsequent step of methane production and that sweet sorghum could be an ideal substrate for a combined gaseous biofuels production (Georgia et al., 2008).

Unseasonal rains beginning in 1995 damaged the maize and sorghum crops harvested in some villages of the Deccan plateau in India. Those grains' human consumption has caused a food-borne disease outbreak characterized by abdominal pain, borborygmi, and diarrhea. The disease was self-limiting. Diarrhea was reproduced in day old cockerels fed contaminated grains from affected households. *Fusarium* sp. was the dominant mycoflora in all 20 sorghum and 12 maize samples collected from affected households. The fumonisin B1 content ranged from 0.14–7.8 mg/kg and 0.25–64.7 mg/kg in sorghum and maize samples, respectively. After harvest, the grains left in the field had higher water activity, which resulted in production of high levels of fumonisin B$_1$. A consumption of such grains by humans caused disease outbreak (Ramesh et al., 1997).

1.6 CONCLUSIONS

Sorghum is an excellent source of protein and starch like other cereal grains. Sorghum can also be processed into different products like flour, starch, grits, and flakes and a large variety of industrial products. However, further research needs to be focused on the utility of sorghum and its products.

KEYWORDS

- cinnamyl alcohol dehydrogenase
- fertile tiller number
- photosynthetically active radiation
- sorghum
- thermal time
- total leaf area per plant

REFERENCES

Aditi, S. S., Naresh, K., & Aggarwal, P. K., (2010). Assessment on vulnerability of sorghum to climate change in India. *Agriculture, Ecosystems, and Environment, 138*, 160–169.

Carberry, P. S., Hammer, G. L., & Muchow, R. C., (1993). Modeling genotypic and environmental control of leaf area dynamics in grain sorghum: III. Senescence and prediction of green leaf area. *Field Crops Research, 33*, 329–351.

Carberry, P. S., Muchow, R. C., & Hammer, G. L., (1993). Modeling genotypic and environmental control of leaf area dynamics in grain sorghum: II. Individual leaf level. *Field Crops Research*, 311–328.

Carolina, C., Luigi, M., Nicola, C., Cristina, B., Luigi, D. G., Domenica, R., Massardo, P., et al., (2007). Celiac disease: *In vitro* and *in vivo* safety and palatability of wheat-free sorghum food products. *Clinical Nutrition, 26*, 799–811.

Edwards, D. N., (1996). Sorghum, beer, and Kushite society. *Journal Norwegian Archaeological Review, 29*, 65–77.

Farré, I., & Faci, J. M., (2006). Comparative response of maize (*Zea mays* L.) and sorghum (*Sorghum bicolor* L. Moench) to deficit irrigation in a Mediterranean environment. *Agricultural Water Management, 83*, 135–143. https://doi.org/10.1016/j.agwat.2005.11.001.

Geoffrey, C., (1994). Fermented foods: A world perspective. *Food Research International, 27*, 253–257.

Georgia, A., Hariklia, N. G., Ioannis, V. S., Angelopoulos, K., & Gerasimos, L., (2008). Biofuels generation from sweet sorghum: Fermentative hydrogen production and anaerobic digestion of the remaining biomass. *Bioresource Technology, 99*, 110–119.

Hammer, G. L., & Muchow, R. C., (1994). Assessing climatic risk to sorghum production in water-limited subtropical environments: I. Development and testing of a simulation model. *Field Crops Research, 36*, 221–234.

Hammer, G. L., & Vanderlip, R. L., (1989). Genotype-by-environment interaction in grain sorghum: I. Effects of temperature on radiation use efficiency. *Crop Science, 21*, 370–376. doi: 10.2135/cropsci1989.0011183X002900020028x.

Hammer, G. L., Carberry, P. S., & Muchow, R. C., (1989). Modeling genotypic and environmental control of leaf area dynamics in grain sorghum: I. Whole plant level. *Field Crops Research, 33*, 293–310.

Hariprasanna, K., Sujay, R., Gomashe, S. S., & Ganapathy, K. N., (2014). Changes in area, yield gains and yield stability of sorghum in major sorghum producing countries, 1970 to 2009. *Crop Science, 54*, 20–29.

Haussmann, B. I. G., Fred, R. E., Weltzien-Rattunde, H., Traoré, K., & Brocke, H. K. P. P. S. C., (2012). Breeding strategies for adaptation of pearl millet and sorghum to climate variability and change in West Africa. *Journal of Agronomy, 5*, 209–218.

Jean-Paul, C., Patricia, M. K., & Peter, M., (1991). On the role of information in decision-making: The case of sorghum yield in Burkina Faso. *Journal of Development Economics, 35*, 261–280.

Jordan, D., Cruickshank, A. W., Borrell, A. K., & Henzell, R. G., (2012). The relationship between the stay-green trait and grain yield in elite sorghum hybrids grown in a range of environments. *Crop Science, 52, 1153–1161.*

Lynn, W., (2006). Worldwide commercial development of bioenergy with a focus on energy crop-based projects. *Biomass and Bioenergy, 30*, 706–714.

Mamoudou, H. D., Gruppen, H., Alfred, T., & Alphons, G. J., (2006). Sorghum grain as human food in Africa: Relevance of content of starch and amylase activities. *African Journal of Biotechnology, 5,* 384–395.

Mohammad, S., & Francis, C. A., (1984). Association of weather variables with genotype × environment interactions in grain sorghum. *Crop Science, 24, 13–16.*

Mukesh, J. P. V., Vara, P., Kenneth, J. B., & Prem, S. C., (2007). Effects of season-long high temperature growth conditions on sugar-to-starch metabolism in developing microspores of grain sorghum (*Sorghum bicolor* L. Moench). *Planta, 22*, 77–79.

Mukesh, J., Chourey, P. S., Kenneth, J. B., & Leon, H. A. Jr., (2010). Short-term high temperature growth conditions during vegetative-to-reproductive phase transition irreversibly compromise cell wall invertase-mediated sucrose catalysis and microspore meiosis in grain sorghum (*Sorghum bicolor*). *Journal of Plant Physiology*, 578–582.

Nicholls, N., (1985). Use of the southern oscillation to predict Australian sorghum yield. *Agricultural and Forest Meteorology, 38,* 8–18.

Normell, J. D., Mesa-Stone, Sajid, A., & Scott, R., (2010). Sorghum proteins: The concentration, isolation, modification, and food applications of kafirins. *Food Science, 75*, 103–113.

Ola, T. W., Mark, A. O., Leo, O., Trygve, B., Hari, U., Siri, B., Siri, D. K. K., et al., (2014). Ethnolinguistic structuring of sorghum genetic diversity in Africa and the role of local seed systems. *PNAS, 11*, 14100–14105.

Palmer, G. H., (1992). Sorghum-food, beverage, and brewing potentials. *Process Biochemistry, 27*, 145–153.

Paul, E. A., & Norman, C. E., (1996). Crop-to-weed gene flow in the genus *Sorghum* (Poaceae): Spontaneous interspecific hybridization between Johnson grass, *Sorghum halepense*, and crop sorghum, *S. bicolor. American Journal of Botany, 83,* 1153–1159.

Paul, W. U., & Louis, B. P., (1999). Factors related to dryland grain sorghum yield increases: 1939 through 1997. *Agronomy Journal Abstract, 91*, 70–79.

Prasad, P. V. V., Boote, K. J., & Allen, L. H., (2011). Longevity and temperature response of pollen as affected by elevated growth temperature and carbon dioxide in peanut and grain sorghum. *Environmental and Experimental Botany, 70*, 51–57.

Prasad, P. V. V., Pisipat, S. R., Mutava, R. N., & Tuinstra, M. R., (2008). Sensitivity of grain sorghum to high temperature stress during reproductive development. *Crop Science Abstract-Crop Physiology and Metabolism, 48, 1911–19.*

Ramesh, V. B., Prathapkumar, H. S., Rao, P., & Sudershan, A., (1997). Foodborne disease outbreak due to the consumption of moldy sorghum and maize containing fumonisin mycotoxins. *Journal of Toxicology: Clinical Toxicology, 35*, 249–255.

Ramirez-Villegas, J., Jarvis, P., & Läderach, (2013). Empirical approaches for assessing impacts of climate change on agriculture: The EcoCrop model and a case study with grain sorghum. *Agricultural and Forest Meteorology, 170*, 67–78.

Russell, E. S., (2003). Taxonomy of sarga, sorghum, and vacoparis (Poaceae: Andropogoneae). *Australian Systematic Botany, 16*, 279–299.

Sanaa, R., El-Sayed, M., Abdel-Aal., & Maher, N., (2006). Antioxidant activity and nutrient composition of selected cereals for food use. *Food Chemistry, 98*, 32–38.

Scott, C. C., (2008). Use of crop models to understand genotype by environment interactions for drought in real world and simulated plant breeding trials. *Euphytica, 161*, 195–208.

Talor, J. R. N., (2011). *Importance of Sorghum in AFRICA.* Department of Food Science, University of Pretoria.

Tuinstra, M. R., & Wedel, J., (2000). Estimation of pollen viability in grain sorghum. *Crop Science, 40*, 230–238.

Vara, P. P. V., & Maduraimuthu, D., (2011). High night temperature decreases leaf photosynthesis and pollen function in grain sorghum. *Functional Plant Biology, 38*, 993–1003.

William, C. M., Guillermo, L. Jr., De Roma, Ñ., Robert, P. P., & George, G. G., (1981). Protein quality and digestibility of sorghum in preschool children: Balance studies and plasma free amino acids. *The Journal of Nutrition, 111*, 1928–1936.

William, L. R., Brent, B., & John, E. M., (2007). Designing sorghum as a dedicated bioenergy feedstock. *Biofuels, Bioproducts and Biorefining, 1*, 147–157.

CHAPTER 2

World Sorghum Production and Factors Affecting Production

ABSTRACT

This chapter presents worldwide sorghum production and application of its products in modern times. It also discusses climatic, abiotic, and biotic factors affecting crop production in different countries.

2.1 SORGHUM PRODUCTION WORLDWIDE

At a time of high petroleum prices and fuel shortages a study was encouraged by the special interest in sugar crops, as probable renewable resources that would supplement non-renewable fossil resources. The study was carried out at eight locations in the continental USA and at one location in Hawaii, where four to six sweet sorghum cultivars were assessed for fermentable sugar production potential. Data was recorded for biomass yield, percent lignin, percent cellulose, stalk sugar yields, and other agronomic characters. In the continental USA total sugar yield varied from 4 Mg ha^{-1} to 10.7 Mg ha^{-1} and up to 12 Mg ha^{-1} at the Hawaiian location. Consequently, hypothetical ethanol production in the continental USA varied from 2129 L ha^{-1} to 5696 L ha^{-1}. The research findings revealed that the sweet sorghum is far more extensively adapted than was expected for a plant of tropical origin and without doubt has the potential to supply a good source of fermentable carbohydrates through a wide geographic area (Smith et al., 1986).

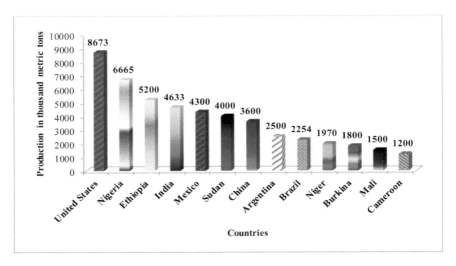

FIGURE 2.1   Sorghum production worldwide in 2019/2020, by leading countries (in 1,000 metric tons).
Source: United States Department of Agriculture (2020).

2.2   APPLICATIONS OF SORGHUM PRODUCTS IN MODERN TIMES

Sweet sorghum [*Sorghum bicolor* (L.) Moench] is well known as a potential feedstock for the production of ethanol around the world. Here, the sweet sorghum and maize (*Zea mays* L.) cultivars were compared for their potential to be a substitute for ethanol production in the northern Corn Belt. Thirteen sweet sorghum cultivars and an adapted maize hybrid were cultivated on clay loam soils. The fermentable carbohydrate content, or Brix (°B), of sorghum stalk sap was assessed using a refractometer. Ethanol yield from sweet sorghum was estimated producing 14.7 pounds fermentable carbohydrate per gallon ethanol and from maize harvested grain by producing 22.4 lb maize grain per gallon of ethanol. The sorghum cultivars Keller, Dale, and Smith yielded higher ethanol than maize and were superior to most other sorghum cultivars or hybrids. However, the sweet sorghum cultivars with the highest fermentable carbohydrate yields were found to be lodging susceptible. Thus, sweet sorghum seems to be a feasible alternative source of ethanol compared with maize for the northern Corn Belt, but the high fermentable carbohydrate cultivars with lodging resistance need to be developed (Daniel et al., 2013).

The increasing cost of energy and limited oil and gas reserves has generated a demand to develop alternative fuels from renewable sources. Presently, the production of renewable transportation fuel is ethanol based. However, ethanol production is based on sugar/starch, the use of these carbohydrates is reduced; as they are the source of food and feed also. The necessity to produce a large and sustainable supply of biomass to make biofuels generation from lignocellulose profitable will need the development of crops grown exclusively for bioenergy production. There are a number of diverse species used as dedicated bioenergy crops, and for several reasons; it is expected that sorghum (*Sorghum bicolor* L. Moench) will be one of these species. Sorghum is a highly productive, drought-tolerant species with a history of improved lignocellulose, sugar, and starch production. In view of this history and the present genetic improvement infrastructure accessible for the species, it is reasonable to assume that sorghum hybrids for dedicated bioenergy production can be developed in the near-term future and will be cultivated and used for bioenergy production (William et al., 2007).

Sweet sorghum contains a sugar-rich stalk, almost like sugarcane. Along with wider adaptability it has rapid growth, high sugar accumulation, and biomass production potential. Water availability is a major constraint in the cultivation of sugarcane. Under such situations sweet sorghum would be a potential substitute for sugarcane, as it can be cultivated with less irrigation and rainfall compared to sugarcane. The sweet sorghum juice contains sugar content varying from 16 to 23% Brix. It has a great potential for jaggery, syrup, and most notably fuel alcohol production. After extraction of juice from sweet sorghum, the stillage can be used for co-generation of power (Reddy et al., 2005).

It is estimated that the consumption of bioethanol can reduce the greenhouse gas emission as biofuel. Owing to its higher tolerance to salt and drought compared to sugarcane and corn, sweet sorghum is considered the best bioethanol source under hot and dry climatic conditions. However, the high carbohydrate content of sweet sorghum stalk is comparable to sugarcane; it requires much lower water and fertilizer than sugarcane. Further, sweet sorghum is tolerant to salinity and its high fermentable sugar content makes it the best choice for fermentation to ethanol. Thus, the study indicated that the problems of increased greenhouse gases and gasoline imports can be solved by using sweet sorghum for the production of biofuel in hot and dry countries (Almodares and Hadi, 2009).

The reduced usage of non-renewable fossil energy reserves and the increased environmental concerns drive interest in bioethanol consumption as an automotive fuel. In this study, ethanol's production from the sugar extract of sweet sorghum juice and from the hemicellulose and cellulose in the leftover sorghum bagasse against selling the sugar from the juice or burning the bagasse to make electricity was examined. Generally, the making of ethanol from the hemicellulose and cellulose in bagasse was more appropriate than burning it to make power, but the comparative benefits of ethanol or sugar production from the juice was very sensitive to the sugar price in China. The process economics and opportunity costs analysis confirmed these results. However, total ethanol production from bagasse of sweet sorghum appears to be very promising, other agricultural crop residues such as corn stover and rice hulls would expect to supply a more attractive feedstock for ethanol production in the medium and long term because of their widespread availability in North China (Gnansounou et al., 2005).

An experiment was conducted on the use of sweet sorghum biomass as a source for hydrogen and methane. The fermentative hydrogen was produced from the sweet sorghum's sugars at different hydraulic retention times (HRT). The highest hydrogen yield, i.e., 10.4 1 H_2/kg sweet sorghum was noted at the HRT of 12 h, whereas the maximum hydrogen production rate of 2550 ml H_2/d was noted at the HRT of 6 h. It was found that the hydrogenogenic reactor effluent yields about 29 1 CH_4/kg of sweet sorghum, which is an ideal substrate for methane production. After the extraction process, 78 1 CH_4/kg was obtained by the anaerobic digestion of the solid residues. This study unveiled that production of biohydrogen and a succeeding step of methane production can be coupled very efficiently and the sweet sorghum could be the best substrate for a combined production of gaseous biofuels (Georgia et al., 2008).

The blending of ethanol in automotive fuels to get clean exhaust and fuel sufficiency increased its demand significantly. The high cultivation cost of sugarcane, sensitive molasses rates, and instabilities in the ethanol price led to a search for an alternate source of ethanol. Owing to its fast growth rate and early maturity, high water use efficiency (WUE), low fertilizer requirement, high total value, and wide adaptability, the sweet sorghum is regarded as a potential raw material for fuel-grade ethanol production. Further, the production of ethanol from sweet sorghum could decrease India's dependency on foreign oil and decrease the environmental threat caused by fossil fuels (Prasad et al., 2007).

An investigation was undertaken to evaluate sweet sorghum cultivars for the variations in biomass, carbohydrates, and ethanol yield (CEY) from anthesis to 40 days after anthesis (DAA). Five sweet sorghum cultivars with crop duration of 111–165 days were grown in Beijing (39°56′N, 116°20′E). An increment in aboveground dry weight (AGDW) and total soluble sugar yield (TSSY, 1.3–10.5 t ha^{-1}) was observed with time after anthesis. The yields of cellulose and hemicellulose vary from 1.6 and 6.6 t ha^{-1} from anthesis to 40 DAA. The stem was found to be the major sink for soluble sugar, with 79.4–94.6% of TSSY. After anthesis an increment in total CEY from the carbohydrates was recorded with time and crop cycle length. The TSSY, cellulose, hemicellulose, and grain yield, and CEY were higher in hybrids when compared with the inbred cultivars with the same crop cycle length. Therefore, it is concluded that both harvest time and genotype exhibit a significant effect on biomass, carbohydrates yield, and CEY. Further, the study recommended that the harvesting of sweet sorghum upon the early maturity of the cultivars from around 20 DAA will give rise to a harvest period of around two months till grain maturity of the late cultivars for ethanol production in North China (Ya et al., 2009).

Sorghum contains various phytochemicals such as tannins, phenolic acids, anthocyanins, phytosterols, and policosanols. These phytochemicals may have a significant effect on human health. Sorghum fractions have high antioxidant activity *in vitro* when compared to other cereals or fruits. These fractions may provide similar health benefits normally provided by fruits. Existing epidemiological evidences suggested that consumption of sorghum decreases the risk of some types of cancer in humans compared to other cereals. This may be partly due to the high concentration of phytochemicals in sorghum. It was widely reported that the tannins in sorghum reduces caloric availability and hence weight gain in animals. This property could help in reducing obesity in humans. Further, sorghum phytochemicals promote cardiovascular health in animals. However, such properties have not been stated in humans and need examination, as cardiovascular disease is a foremost killer disease in the developed world at present. Here, accessible literature on sorghum phytochemicals and their potential to prevent common nutrition-related diseases such as cancer, cardiovascular disease, and obesity were reviewed (Dykes and Linda, 2019).

2.3   FACTORS AFFECTING SORGHUM PRODUCTION

The dynamics of food production need to be understood to improve food security. This is most significant in regions that depend on subsistence agriculture with little adaptive capacity to climate change. Sorghum plays a pivotal role in food security in some of the poorest sections of the world. So in this paper, an attempt was made to identify the key factors influencing sorghum production in three major production regions by reviewing the literature. Sorghum production was found to be affected by ten major factors: climate change, population growth/economic development, non-food demand, agricultural inputs, demand for other crops, agricultural resources scarcity, biodiversity, cultural influence, price, and armed conflict. This indicates that several factors affect sorghum production simultaneously and the effect of each factor is highly influenced by the extent and certainty of one or more other factors. These factors vary in significance and amount with reference to geography. In general, improved agricultural inputs, population growth/economic development, and climate change have considerable influence on sorghum production. However, it is expected that local dynamics go ahead of these broad trends and more comprehensive, locally focused studies are required for actionable planning purposes (Clara et al., 2019).

An experiment was conducted to study the effect of the conventional agricultural inputs on sorghum production levels in the Gezira scheme. Further, the key factors behind the tenant's technical inefficiency and their effects on food security of the household level were studied. In the Gezira Scheme, 100 tenants samples were utilized to estimate the stochastic frontier production function. The results revealed that the levels of sorghum production were significantly and positively affected by credit, capital, hired labor, fertilizer, and irrigation. Sorghum area had a negative and significant effect on levels of sorghum production. The average technical efficiency for sorghum production was 67%. This indicates that the effective use of the tenant's available resources could increase the sorghum yield. Size of holding, education level, tenants experience, household size, contact with extension agents and farm location are important in elucidating tenants' technical inefficiency (Adam et al., 2005).

Sorghum (*Sorghum bicolor* (L.) Moench) is a major cereal crop grown in semi-arid tropics. In Kenya, it is cultivated in the drought-prone marginal agricultural areas. Sorghum is well adapted to drought and flood-prone areas because of its C_4 photosynthetic nature, widespread root system, waxy leaves, and ability to cease growth during drought. Sorghum is closely related to maize in utilization and could be used as its substitute in drought-prone areas. In spite of its suitability in the semiarid areas, the area under sorghum production is still low and farmers get low yields in eastern Kenya. The majority of the farmers still cultivate maize which is subjected to frequent crop failures. This study was undertaken was to collect information on socio-economic factors influencing sorghum production and the sorghum farming system utilized by the farmers in the region, landraces grown by farmers, source of seed, preferred character-istics, maturity period, cultural practices, pre, and post-harvest handling, use, and limitations in sorghum production in lower eastern Kenya region. The studied parameters were expressed as percentages and bar graphs constructed. Analysis of variance (ANOVA) was done and least signifi-cant differences were used to separate means at 0.05 level of confidence. It was found the informal systems provide most of the sorghum used as food and seed for planting. Farmers use low inputs due to low income. The Eastern region's main occupation is agriculture and crop productivity could be increased by using locally available germplasm. They found that a diversity of sorghum landraces maintained by farmers in eastern Kenya. Owing to its high iron and zinc content, sorghum grain could be used to decrease micronutrient malnutrition. However, its production is low because of limitations such as deficit of income to purchase fertilizer and chemicals, insufficient quality seed, susceptibility to pests and diseases which results in low yields (Muui et al., 2013).

Low adoption of agricultural technology is the foremost reason for low farm productivity and a high incidence of poverty and food insecurity in sub-Saharan countries including Tanzania. Here, the factors influencing the adoption of improved sorghum varieties were examined. In northern and central Tanzania 822 sample households were selected randomly. A multiple-hurdle Tobit model was employed to determine the factors influencing adoption after controlling for both capital and information limitations. Further, they used t-distributed stochastic neighbor embed-ding to cluster farmers into homogeneous groups. This method reduces the dimensionality by maintaining the topology of the dataset, which improves the clustering accuracy. Results unveiled that radio and other mass media channels that create awareness increase adoption among farmers reducing capital constraints. This method has a high impact on farmers who do not have fundamental resources such as land and capital, and subsidies. Other farmers just need assurance on the performance of improved sorghum varieties. Field days, on-farm trials, and demonstration plots could be helpful to support these farmers. However, a tailored support system is required to sustain investment in both quantity and quality of services. Therefore, there is a need to develop a pluralistic research and extension systems that promote the usage of information technologies and community-based organizations to reach particular groups of farmers (Aloyce et al., 2018).

The factors that affect the sugar yield of sweet sorghum were determined at temperature zone locations. In this study, four sweet sorghum cultivars were assessed for fermentable sugar production potential under irrigated and non-irrigated conditions. The cultivars were applied with 0, 84, and 186 kg ha^{-1} of nitrogen fertilizer at two temperature zone locations (40·8° and 42°N latitude). The average yield of ethanol (EtOH) ranged from 3100 to 5235 liters ha^{-1}. However, irrigated, and non-irrigated conditions did not show any significant differences for total sugar yield and theoretical EtOH, the gross green weight (89·9 Mg ha^{-1}) was higher under irrigated condition than non-irrigated condition. The nitrogen fertilizer showed an insignificant effect on increasing fermentable sugar production. Therefore, the study results suggested the sweet sorghum as a potential substitute energy crop, capable of insulating ethanol prices from shifts in corn prices (Smith, 1993).

A study was conducted to determine the optimum harvesting time, sugar production, biomass composition and ethanol yields of sweet and forage sorghum in Northeast Mexico. The juices from the sweet sorghum cultivars Rio, M81E and Della and forage sorghum cultivars Tato and Thor were characterized for the sugar composition, free amino nitrogen (FAN) and phenolics and then yeast (*Saccharomyces cerevisiae*)-fermented into ethanol. Different sweet juice volumes, total sugars, and optimum harvesting times were observed in the cultivars. The results revealed that fructose and sucrose were the most predominant sugar in raw juices and FAN concentration ranged from 19 to 36 mg L^{-1}. This indicated that a nitrogen supplement was needed for sufficient fermentation. However, after 18 hours of fermentation, significant differences were not observed in productivities among cultivars, but the ethanol yield was higher in sweet sorghum cultivars compared to the two forage sorghum cultivars (Davila-Gomez et al., 2011).

Sweet sorghum [*Sorghum bicolor* (L.) Moench] production can be optimized by following good management practices. The effects of N rate (45, 90, 135, and 180 kg N ha^{-1}) and top removal (at boot stage, anthesis, and none) on biomass, Brix, assessed sugar yield, and N and P recovery of sweet sorghum cv. M-81E were studied in Florida at two locations varying in soil type. At both locations, N fertilization rate did not show any effect and an average dry biomass yield was 17.7 Mg ha^{-1}. Average Brix values ranged from 131 to 151 mg g^{-1} and had a negative correlation with N rate. Removal of top at boot stage or anthesis provided greater Brix values and 13% higher sugar yields at both locations. Plant N recovery ranged from 78 to 166 kg N ha^{-1} two-thirds of which was found in leaf and grain tissues. The N recovery exhibited a positive association with the N rate. On the basis of yield and nutrient recovery responses 90 to 110 kg N ha^{-1} and 15 to 20 kg P ha^{-1} were estimated as optimum nutrient doses. However, higher N recovery was recorded at higher N fertilization; gain in sugar yield was little to modest. Therefore, to improve sugar yields of sweet sorghum for the production of biofuels and bio-based products appropriate nutrient and harvest management is required. Further, the management practices of sweet sorghum for bioenergy production, particularly relating to the usage of leaf and grain tissue should be refined (John et al., 2012).

2.4   CONCLUSIONS

Sorghum is the fifth most vital grain crop used as a food and feed crop in the agroecosystem of the world. Sorghum might have attained a more noticeable status as an international cash crop decades before. Ten major factors such as climate change, population growth/economic development, non-food demand, agricultural inputs, demand for other crops, agricultural resources scarcity, biodiversity, cultural influence, price, and armed conflict affects sorghum production. Consequently, there is a need to develop diverse research and extension systems to improve sorghum production.

KEYWORDS

- abiotic factors
- aboveground dry weight
- factors affecting yield
- free amino nitrogen
- hydraulic retention times
- sorghum

REFERENCES

Adam, E. A., Kuhlmann, F., Mau, M., Elobeid, H. A., & Elamin, E. M., (2005). Analysis of factors affecting sorghum production in the Gezira scheme-Sudan and implications on the household food security. *Conference on International Agricultural Research for Development.* Stuttgart-Hohenheim.

Almodares, A., & Hadi, M. R., (2009). Production of bioethanol from sweet sorghum: A review. *African Journal of Agricultural Research, 4,* 772–781.

Aloyce, R. K., Kizito, M., Theresia, L. G., Frida, M. M., & Mary, M., (2018). Factors affecting adoption of improved sorghum varieties in Tanzania under information and capital constraints. *Agricultural and Food Economics, 6,* 1–21.

Clara, W. M., Silvia, S., Kofi, A., & Guangxing, W., (2019). A regional comparison of factors affecting global sorghum production: The case of North America, Asia, and Africa's Sahel. *Sustainability, 11,* 1–18.

Daniel, H. P., William, E. L., Brian, K. K., & Thomas, R. H., (2013). Comparison of sweet sorghum cultivars and maize for ethanol production. *Journal of Production Agriculture, 4,* 121–130.

Davila-Gomez, F. J., Chuck-Hernandez, C., Perez-Carrillo, E., Rooney, W. L., & Serna-Saldivar, S. O., (2011). Evaluation of bioethanol production from five different varieties of sweet and forage sorghums (*Sorghum bicolor* (L) Moench). *Industrial Crops and Products, 33,* 611–616.

Dykes, L., (2019). Sorghum phytochemicals and their potential impact on human health. *Methods in Molecular Biology, 3,* 78–86.

Georgia, A., Hariklia, N. G., Ioannis, V. S., Angelopoulos, K., & Gerasimos, L., (2008). Biofuels generation from sweet sorghum: Fermentative hydrogen production and anaerobic digestion of the remaining biomass. *Bioresource Technology, 99,* 110–119.

Gnansounou, E., Dauriat, A., & Wyman, C. E., (2005). Refining sweet sorghum to ethanol and sugar: Economic trade-offs in the context of North China. *Bioresource Technology, 96,* 985–1002.

John, E. E., Kenneth, R. W., & Lynn, E. S., (2012). Sweet sorghum production for biofuel in the southeastern USA through nitrogen fertilization and top removal. *Bioenergy Research, 5,* 86–94.

Muui, C. W., Muasya, R. M., & Kirubi, D. T., (2013). Baseline survey on factors affecting sorghum production and use in Eastern Kenya. *African Journal of Food Agriculture Nutrition and Development, 13*, 7340–7359.

Prasad, S., Anoop, S., Jain, N., & Joshi, H. C., (2007). Ethanol production from sweet sorghum syrup for utilization as automotive fuel in India. *Energy Fuels, 21*, 2415–2420.

Reddy, B. V. S., Ramesh, S., Reddy, P. S., Ramaiah, B., Salimath, M., & Kachapur, R., (2005). Sweet sorghum: A potential alternate raw material for bio-ethanol and bio-energy. *International Sorghum and Millets Newsletter, 46*, 79–86.

Smith, A., Bagby, M. O., & Lewellan, R. T., (1986). Sweet sorghum for fermentable sugar production. *Crop Science, 27, 788–793*.

Smith, G. A., Buxton, D. R., Smith, G. A., & Buxton, D. R., (1993). Temperate zone sweet sorghum ethanol production potential. *Bioresource Technology, 43*, 71–75.

William, L. R., Jürg, B., Brent, B., & John, E. M., (2007). Designing sorghum as a dedicated bioenergy feedstock. *Biofuels, Bioproducts and Biorefining, 1*, 147–157.

Ya, L. Z., Abdughani, D., Yosef, S., Xin, W., Amarjan, O., & Guang, H. X., (2009). Biomass yield and changes in the chemical composition of sweet sorghum cultivars grown for biofuel. *Field Crops, 111*, 55–64.

CHAPTER 3

Origin, Evolution, and Domestication of Sorghum

ABSTRACT

This chapter deals with the origin, evolutions, and domestication of sorghum. This discusses research advances on the origin, evolution, and domestication of sorghum-based on phylogeny, cytology, and molecular levels. Sufficient research advances have been made on these important aspects.

3.1 REGIONAL HISTORY OF SORGHUM

Sorghum is the fifth most important cereal crop used as a food and feed crop in the agroecosystem of the world on the basis of tonnage of its grain production next to maize, wheat, rice, and barley grain crops. It is strangely tolerant of low input levels. This is a common characteristic feature seen in many areas of Northeast Africa and also some of the regions in the US Southern plains. These areas specifically receive a very meager amount of rainfall which fails to meet the growing requirements of other grain crops. With an increase in demand for limited sources of freshwater, an increase in the usage of marginal lands for agricultural production and variability's in the trends of global climate, to meet the needs of the world's increasing population growth, some of the dry land crops as sorghum will be more valuable as it is one of the low input requiring crop.

3.2 ECOTYPES AND CULTIVARS OF SORGHUM

De Wet (1978) stated that the genus *Sorghum* Moench is subdivided into sections *Chaeotosorghum, Heterosorghum, Parasorghum, Stiposorghum,* and *Sorghum*. Section *Sorghum* has two rhizomatous species, *S. halepense* (L.) Pers. ($2n$ = 40) and *S. propinquum* (Kunth) Hitchcock ($2n$ = 20), in addition to the annual *S. bicolor* (L.) Moench ($2n$ = 20). *Sorghum bicolor* is categorized into subspecies *bicolor* which includes all domesticated grain sorghums, subspecies *drummondii* (Steud.) de Wet comb.nov. which include stabilized derivatives of hybridization among grain sorghums and their closest wild relatives and subspecies *arundinaceum* (Desv.) De Wet and Harlan (1965) mentioned that it includes the wild progenitors of grain sorghums. They recognized four ecotypes of subspecies *arundinaceum*: race *aethiopicum* of the arid African Sahel, race *virgatum* of northeastern Africa, race *arundinaceum* of the African tropical forest, and race *verticilliflorum* of the African Savanna. Others recognized that grain sorghums are usually divided among five basic races, bicolor, caudatum, durra, guinea, and kafir, and ten hybrid races that each combines characteristics of at least two of these basic races. Grain sorghum races are morphologically dissimilar, and they preserve their unity of type through spatial and ethnological isolation.

Evans et al. (2010) described the eco-geographical distribution of wild, weedy, and cultivated *Sorghum bicolor* (L.) Moench in Kenya and its inferences for conservation and also about the crop-to-wild gene flow. They mentioned that the overlaps in ecological and geographical distributions determine possible gene flow between a crop and its wild relatives. Therefore, the ecogeographical databases are the key tools for the sustainable management of genetic resources. They performed *in situ* collections of wild, weedy, and cultivated sorghum. For each sampled wild sorghum plant qualitative and quantitative morphological traits measurements were taken. Farmers' knowledge about the management of sorghum varieties and autecology of wild sorghum was also obtained. Cluster analysis supported several wild sorghum morphotypes that might resemble at least three of the five ecotypes identified in Africa. Intermediate forms between wild and cultivated sorghum belonging to the *S. bicolor* ssp. *drummondii* were often found in predominantly sorghum growing areas. Crop-wild gene flow in sorghum is probable to occur in many agroecosystems of Kenya.

Vegetative dispersal by rhizomes (underground stems) and seed dispersal by disarticulation of the mature inflorescence (shattering) cause "Johnsongrass" [*Sorghum halepense* (L.) Pers.], ranks it as one of the world's most noxious weed (Holm et al., 1977). It is an interspecific hybrid of *Sorghum bicolor* and *S. propinquum*. To this interspecific hybrid the rhizamotousness is mostly contributed by *S. propinquum*. Because of their capacity to readily cross, the progeny thus obtained from the above crosses provides a system that enables in the dissection of the genetic basis of rhizomatousness (Paterson et al., 1995).

Most of the archaeological and historical evidences have indicated that the genus Sorghum owes its distribution from the Old World and has extended over its continuous distribution from regions of Southern Africa to Subtropical Australia. However, the studies have shown that only after Australia's colonization by the Europeans that this cultivated weed of sorghum complex has reached Australia (De Wet and Huckaray, 1996). De Wet and Huckaray (1996) have mentioned that subspecies of sorghum viz., *Sorghum bicolor* subsp. halepense had its wide distribution in the areas of the Mediterranean regions and had further exhibited its distribution to the islands of the South East Asian region.

Harlan and De Wet (1965) mentioned that *Sorghum bicolor* subsp. *halepense* is often encountered in areas more prone to disturbances by man. Particularly, it has been prevalent in those man-disturbed localities present in the Mediterranean and Tropics. De Wet (1967) found two morphologically diverse complexes in this subspecies one with a Mediterranean type and the other a tropical ecotype. He further mentioned that the Mediterranean ecotype consisted of petite plants with more narrowed leaves. It has extended eastward from across Asia Minor. It reached the mountains of the West Pakistan. In the mountains of West Pakistan, this weed was subjected to the replacement by the robust tropical ecotype of the subspecies of halepense. These robust tropical ecotypes had broad leaves. They further extended to the parts of Western India. Similarly, another subspecies of sorghum, i.e., which comprises most of the cultivated sorghums, companion weeds and some semiwild relatives was introduced into Punjab from African regions during the historical period (De Wet and Huckaray, 1996). Similarly, other weed species of sorghum that had extended eastward to Somaliland and the Nile valley (De Wet and Huckaray, 1996).

3.3   ORIGIN OF SORGHUM

The importance of a quantitative study in ascertaining associations within *S. bicolor* was explained by Liang and Casady (1966). Their research was based on plants cultivated for many generations in various nurseries. The small genome of sorghum for a long time has been an attractive model to advance the knowledge towards the structure, function, and evolution of many of the cereal genomes. Sorghum is a representative of tropical grasses, having C_4 photosynthesis. It uses its complex biochemical and morphological specializations for improving the assimilation of carbon at high temperatures (HTs). Sorghum is much more intimately related than rice to a lot of the most important cereal crops with complex genomes and high levels of gene duplication. Sorghum and Zea (maize, the leading US crop with a farm-gate value of $15–20 billion/y) diverged from a common ancestor ~12 mya (Gaut et al., 1997; Swigonowa et al., 2004).

3.4   EVOLUTION OF SORGHUM

Kimber (2000) discussed the origin, history, technology, and production of sorghum and its early distribution to India and China. In his book, he discussed the state of diffusion of sorghums to Asia, the archaeological evidence, the climatic situations, Archaeological sites of China, early introductions, and evidence.

De Wet and Harlan (1975) mentioned that weeds evolved, and are still evolving, within the man-made habitat in three most important ways: from colonizers during selection in the direction of adaptation to continuous habitat disturbance; as products of hybridization between wild and culti-vated races of domestic species; and through selection in the direction of re-establishing natural seed dispersal mechanisms in neglected domesti-cates. Domesticates show variation from weeds mainly in the extent of dependence on man for survival. They evolved from wild food plants which were brought into cultivation. The process of domestication commenced when the propagation of plants for successive generations was started by man using seed or vegetative propagules. Phenotypic variations related to planting and harvesting are species-specific, and are created by natural selection under cultivated conditions. During the domestication process artificial selection by man was a major cause for subspecific variation in domestic species.

The phylogenetic associations of the genus *Sorghum* and related genera were examined by sequencing the nuclear ribosomal DNA (rDNA) internal transcribed spacer region (ITS) (Sun et al., 1994). In this study, they extracted DNA from 15 *Sorghum* accessions, comprising one accession from each of the sections *Chaetosorghum* and *Heterosorghum*, four accessions from *Parasorghum*, two accessions from *Stiposorghum*, and seven representatives from three species of the section *Sorghum* (one accession from each of *S. propinquum* and *S. halepense*, and five races of *S. bicolor*). They used variable nucleotides to build a strict consensus phylogenetic tree. The studies indicated that *S. propinquum*, *S. halepense* and *S. bicolor* subsp. *arundinaceum* race aethiopicum may be the nearest wild relatives of cultivated sorghum; *Sorghum nitidum* may be the closest $2n = 10$ relative to *S. bicolor*, the sections *Chaetosorghum* and *Heterosorghum* seem closely associated to each other and more strongly associated to the section *Sorghum* than *Parasorghum* and the section *Parasorghum* is not monophyletic. The findings also specified that the genus *Sorghum* is a very ancient and diverse group.

Variations in the DNA content have a key vital role in understanding plant evolution and adaptation which remains as a major biological enigma. A *Sorghum* phylogeny was created on the basis of combined nuclear ITS and chloroplast *ndhF* DNA sequences by James Price et al. (2003). Their findings revealed that chromosome counts ($2n = 10, 20, 30, 40$) were, with few exceptions. New chromosome numbers were found for *S. amplum* ($2n = 30$) and *S. leiocladum* ($2n = 10$). Among the 21 Sorghum species 2C DNA content ranged from 1.27–10.30 pg (8·1-fold). However, among the $2n = 10$ species and $2n = 20$ species 2C DNA content varies 3·6-fold (1·27 pg to 4·60 pg) and 5·8-fold (1·52–8·79 pg), respectively. The $x = 5$ genome size varies over an 8·8-fold range from 0·26 pg to 2·30 pg. Perennial species (6.20 pg) had significantly higher mean 2C DNA content than the annuals (2·92 pg). Among the 21 species studied, a significant difference was not seen in the mean $x = 5$ genome size of annuals (1·15 pg) and perennials (1·29 pg). Statistical analysis of Australian species revealed: (*a*) a significant difference in mean 2C DNA content of annual (2·89 pg) and perennial (7·73 pg) species; (*b*) in perennials mean $x = 5$ genome size (1·66 pg) is significantly greater than that of the annuals (1·09 pg); (*c*) the mean maximum latitude at which perennial species grow (−25·4°) is significantly greater than the mean maximum latitude (−17·6) at which annual species grow. Further, this DNA sequence phylogeny studied by James Price et al. (2005) in 21 species of sorghum, divides *Sorghum* into two lineages, one containing the $2n = 10$ species with large genomes and their polyploid relatives, and the other with the $2n = 20, 40$ species with comparatively small genomes. An evident phylogenetic decrease in genome size seemed to occur in the $2n = 10$ lineage. It was found that the genome size evolution in the genus *Sorghum* may not include a 'one-way ticket to genomic obesity' as has been put forward for the grasses.

Sorghum [*Sorghum bicolor* (L.) Moench] breeders have reported the significance of exotic germplasm and non-cultivated sorghum races as sources of valuable genes for genetic improvement long ago. The genus *Sorghum* consists of 25 species categorized as five sections: *Eu-sorghum*, *Chaetosorghum*, *Heterosorghum*, *Para-sorghum*, and *Stiposorghum*. Species outside the *Eu-sorghum* section are sources of valuable genes for sorghum improvement, including those for insect and disease resistance, but because of the failure of these species to cross with sorghum these have not been used. The success of hybridization and introgression between sorghum and divergent sorghum species depends on the understanding of the biological nature of the incompatibility system(s) that prevent hybridization and/or seed development. George et al. (2004) by utilizing 14 alien *Sorghum* species proved that pollen-pistil incompatibilities were the major causes that prevent hybridization in sorghum. The alien pollen tubes show major inhibition of growth in sorghum pistils and rarely grew beyond the stigma. Pollen tubes of only three species grew into the ovary of sorghum. Fertilization and following embryo development were not common. It seems that the breakdown of the endosperm aborts seeds with developing embryos before maturation.

3.5   DOMESTICATION OF SORGHUM

Almost all cultivated sorghums around the world and a few semi-wild plants are included under the species *Sorghum bicolor*. This species unveils large variations in morphological traits. Snowden (1936, 1966) based on its complexity has subdivided it into cultivated (28) and related wild (28) species. Because of the absence of the genetic barrier between these taxa, has led to an indication that all of these belong to a single species (De Wet, 1967).

Domestication of sorghum began around the waters of the Niger particularly the headwaters by the Mande people (Murdock, 1959). Wrigley (1960) and Baker (1962) have pointed out that the plant domestication of sorghum was most likely introduced from the Near East to the

Ethiopian region and later it has been spread to the regions of West and South Africa and also to the Southeast Africa. Doggett (1965) on the basis of the archaeological evidences has mentioned that its domestication dates back to 3000 B.C.

The information on its antiquity is limited. There is no conclusive archaeological evidence about its cultivation in the prehistoric era. Due to a lacuna in the historical and archaeological data records the scientists have resolved that origins and domestications of sorghum for the present must be based on the studies of comparative morphological features that are mostly associated with its distribution patterns (De Wet and Huckaray, 1996). Selection of the early farmers from among the African wild types which were morphologically distinct and variable with a variety of uses of the wild types of *Sorghum bicolor*, has enabled in its spread over to large regions of the old world. With the migration of man, who has undertaken new hybrid combinations were resulted by the crossings over of geographically isolated cultivars and the wild races which were brought together. These new hybrid combinations were under continuous production. The acceleration in this morphological variation was more pronounced during the past 500 years. In many of the newly settled colonies as most of the cultivated and weed sorghums were brought under cultivation, this has led to the prevalence of the current widespread hybridization to take place (De Wet and Huckaray, 1996).

Some of the studies have shown that during the process of sorghum domestication underwent a loss of genetic variation. A study of allozyme variation among the spontaneous taxa of *Sorghum* section Sorghum was presumed by Morden et al. (1990). Eight plants from 90 accessions of diploid *S. bicolor* (ssp. *arundinaceum* and *drummondii*) and the tetraploids *S. almum* and *S. halepense* were examined for 17 enzyme systems encoded by 30 loci. Significant variations were not seen within and among accessions. Although the differences of variation were higher than that of typical inbreeding species, they found that an average of 3.2 alleles per locus in ssp. *arundinaceum*, with an average estimated heterozygosity for the accessions of 0.034 and total panmictic heterozygosity of 0.154. An analysis of the distribution of genetic variation among accessions of ssp. *arundinaceum* revealed that 26% of the variation takes place within accessions and 74% among accessions. Cultivated sorghum holds much less allozymic variation than ssp. *arundinaceum*, its assumed progenitor. This is reliable with the prediction that cultivated sorghum underwent a loss of genetic variation during domestication. Primarily, cultivated sorghum contains a subset of the allozymes observed in ssp. *arundinaceum*. Principal component analysis unveiled continuous variation among the accessions and geographic regions, with accessions failing to segregate into different clusters. However, accessions of race *virgatum* of ssp. *arundinaceum* took one end of the continuum and were, in that sense, differentiated from the other accessions. Likewise, most accessions of *S. halepense* and *S. almum* occupied the central portion of the continuum. The allozymic data reported was consistent with the assumed origin of *S. halepense* via autopolyploidy or segmental allopolyploidy.

Sally et al. (2007) discussed on domestication to crop improvement of *Sorghum* and *Saccharum* (Andropogoneae). Both sorghum (*Sorghum bicolor*) and sugarcane (*Saccharum officinarum*) are closely related and members of the Andropogoneae tribe in the Poaceae. Both are comparatively recent domesticates and relatively little of the genetic potential of these taxa and their wild relatives have been captured by breeding programs so far. In these review genetic gains made by plant breeders since domestication and the advancement in the categorization of genetic resources and their application in crop improvement for these two associated species were assessed. Recently sorghum genome has been sequenced which provided a great boost to our understanding of the evolution of grass genomes and the prosperity of diversity within *S. bicolor* taxa. Molecular analysis of the *Sorghum* genus has detected new traits, endosperm structure, and composition in close relatives of *S. bicolor*. These might be helpful to enlarge the cultivated gene pool. Mutant populations (including TILLING populations) offer a valuable addition to the genetic resources of this species. The improvement of sorghum can be speeded up by the integration of more distinct germplasm into the domesticated gene pools employing molecular tools and the enhanced knowledge of its genomes.

Michael et al. (2009) in their discussion on the nature of selection during plant domestication has mentioned that it is a wonderful example of plant-animal co-evolution and is a far richer model for exploring evolution than is usually appreciated. Numerous studies have been conducted to find genes linked with domestication, and archaeological work, which presented a clear perception of the dynamics of human cultivation practices during the Neolithic period. Together, these have provided a better understanding of the selective pressures that go with crop domestication. Further, they demonstrated that a synthesis from the twin vantage points of genetics and archaeology can increase our knowledge of the nature of evolutionary selection that accompanies domestication.

Doggett (1965) and De Wet (1967) have discussed in detail about the origins of the cultivated sorghums. They mentioned that it had been first domesticated in Africa (Burkill, 1936). Based on the available morphological data, they suggested that the distribution of sorghum had three independent centers of domestication in Africa. The West African Negroids grew guinea corn, while the Bantu grew kafir corn. Most of the races of *Sorghum bicolor* and durra are cultivated by the Arabic stock people in the African regions (De Wet and Huckaray, 1996).

As sorghum is a characteristic of multiple-origin species, the genetic basis of its domestication and improvement is not known yet. Huanhuan et al. (2019) studied the genetic architecture of traits associated with domestication and improvement of sorghum using F_2 and F_3 populations acquired by crossing *Sorghum virgatum* and domesticated sorghum. It was found that human selection had significantly reformed sorghum through the quantitative trait loci (QTLs) with large genetic effects in the characters of harvest, plant architecture and grain taste, which includes the decrease in shattering, few branches, short plant stature and the exclusion of polyphenols from seed. They selected the enlargement of seed width to increase the yield by gathering small-effect QTLs. The two most important QTLs of plant height (QTI-ph1 and dw1) were constricted to 24.5-kilobase (kb) and 13.9-kb, respectively. DNA diversity analysis and association mapping of *dw1* gene revealed the functional variant (A1361 T) might have originated from the same event not long time ago. The study findings established that corresponding phenotypic variations across diverse species during domestication and improvement may share the identical genetic basis and QTL × QTL interactions may not have a pivotal role in the transformation of characters during sorghum domestication and improvement. Further, the study provided new insights into transgressive segregation and segregation distortion and significantly improved the knowledge of the genetic basis of sorghum domestication and improvement.

3.6 CONCLUSIONS

Numerous studies have been undertaken on origin, evolution, and domestication of sorghum on the archeological, phylogeny, cytological, and molecular levels. Several theories are put forth by different authors on these aspects. This chapter considered the origin of sorghum, the process of evolution, polyploidization, hybridization giving rise to the origin of diverse species and steady process of domestication.

KEYWORDS

- cytology
- internal transcribed spacer
- molecular biology
- phylogeny
- polyploidy
- sorghum

REFERENCES

Baker, H. G., (1962). Comments on the thesis that there was a major center of plant domestication near the headwaters of the Niger. *The Journal of African History, 3*, 229–233.

Burkill, I. H., (1936). The races of sorghum. *Kew Bull.*, 112–119.

De Wet, J. M. J., & Harlan, J. R., (1975). Weeds and domesticates: Evolution in the man-made habitat. *Economic Botany, 29*, 99–108.

De Wet, J. M. J., (1978). Special paper: Systematics and evolution of sorghum sect. sorghum (Gramineae). *American Journal of Botany, 2*, 477–484.

De Wet, J. M., & Huckabay, J. P., (1996). The origin of *Sorghum bicolor*: II. Distribution and domestication. *Evolution, 21*, 787–802.

De Wet, J. M., (1967). The origin of *Sorghum bicolor*: I. Morphological variation. *Bot. Gaz.*

Doggett, H., (1965). The development of cultivated sorghums. In: Sir Joseph, H., (ed.), *Crop Plant Evolution*. Cambridge University Press, London.

Evans, M., Fabrice, S., & Moses, M. E. T., (2010). Ecogeographical distribution of wild, weedy, and cultivated *Sorghum bicolor* (L.) Moench in Kenya: Implications for conservation and crop-to-wild gene flow. *Genetic Resources and Crop Evolution, 57*, 243–253.

Gaut, B. S., Clark, L. G., Wendel, J. F., & Muse, S. V., (1997). Comparisons of the molecular evolutionary process at rbcL and ndhF in the grass family (Poaceae). *Molecular Biology and Evolution, 14*, 769–778.

George, L., Byron, L., William, L., Rooney, Sally, L., Dillon, & James, P. H., (2004). Pollen–pistil interactions result in reproductive isolation between *Sorghum bicolor* and divergent *Sorghum* species. *Crop Science, 45,* 1403–1409.

Harlan, J. R., & De Wet, J. M. J., (1965). Some thoughts about weeds. *Economic Botany, 19,* 16–24.

Holm, L. G., Plucknett, D. L., Pancho, J. V., & Herberger, J. P., (1977). *The World's Worst Weeds: Distribution and Biology.* University Press of Hawaii, Honolulu, Hawaii, USA.

Huanhuan, L., Hangqin, L., Leina, Z., & Zhongwei, L., (2019). Genetic architecture of domestication – and improvement – related traits using a population derived from *Sorghum virgatum* and *Sorghum bicolor. Plant Science, 283,* 135–146.

James, P., Salty, H., Dillon, G., Hodnett, W., Rooney, Larry, R., & Spener, J., (2005). Genome evolution in the genus sorghum (Poaceae). *Annals of Botany, 95,* 219–227.

Kimber, (2000). *Origins of Domesticated Sorghum and its Early Diffusion to India and China-Cited in Sorghum: Origin, History, Technology, and Production.* Books Google.

Michael, D. P., & Dorian, Q., (2009). The nature of selection during plant domestication. *Nature, 457,* 843–848.

Murdock, G. P., (1959). Africa-its people and their cultural history. McGraw-Hill, New York. 456 p.

Paterson, A. H., Schertz, K. F., Lin, Y. R., Liu, S. C., & Chang, Y. L., (1995). The weediness of wild plants: Molecular analysis of genes influencing dispersal and persistence of Johnson grass, *Sorghum halepense* (L.). *Proceedings of the National Academy of Sciences of the United States of America, 92,* 61–67.

Sally, L., Dillon, F. M., Shapter, Robert, J., Henry, G., Cordeiro, L. I. L., & Slade, L., (2007). Domestication to crop improvement: Genetic resources for *Sorghum* and *Saccharum* (Andropogoneae). *Annals of Botany, 5,* 975–989.

Sun, Y., Skinner, D. Z., Liang, G. H., & Hulbert, S. H., (1994). Phylogenetic analysis of *Sorghum* and related taxa using internal transcribed spacers of nuclear ribosomal DNA. *Theoretical and Applied Genetics, 89,* 26–32.

Swigonová, Z., Lai, J., & Ma, J., (2004). Close split of sorghum and maize genome progenitors. *Genome Research, 14,* 1916–1922.

Wrigley, C., (1960). Speculation on the economic prehistory of Africa. *The Journal of African Histrory, 1,* 189–203.

CHAPTER 4

Sorghum Ideotype

ABSTRACT

An ideal plant type helps in the capture and interception of solar radiation, more number of grains productions with good quality and yield. This chapter briefly highlights some of the research activities carried out in the development of ideal plant types of sorghum suitable for different environmental conditions and planting densities.

4.1 ARCHITECTURES OF SORGHUM PLANT

Higher plants exhibit different architectures for the extent of branching, internodal elongation, and shoot determinacy. Researches on the model plants of *Arabidopsis thaliana* and tomato and crop plants like rice and maize have significantly increased our understanding of the molecular genetic bases of plant architecture. The identification of mutants that cause defects in plant architecture and characterization of the corresponding and associated genes will ultimately explain the molecular mechanisms underlying plant architecture. Thus far, the accomplishments made in studying plant architecture have already permitted the optimization of crop plant architecture by molecular design to improve grain productivity (Yonghong and Jiayang, 2008).

The root system determines a plant's ability to remove soil water and its architecture can affect adaptation to water-stress conditions. The architecture can be associated with the root characteristics of young plants to rapidly screen the phenotype. For this improved understanding of root system development is required. So, in this study the morphological and architectural development of sorghum and maize root systems were characterized by (i) observing the origin and initiation of roots at germination, and (ii) observing and measuring the development of root systems in young plants. Three experiments were performed using two maize and four sorghum hybrids. Single seminal (primary) root was produced in sorghum and coleoptile nodal roots appeared at the 4th–5th leaf stage, while 3–7 seminal (primary and scutellum) roots were produced in maize and coleoptile nodal roots appeared at the 2nd leaf stage. The flush angle and average diameter of nodal roots showed genotypic variation. Therefore, they could be taken into account as an appropriate target for large-scale screening of root architecture in breeding populations. In sorghum, such screening would need a small chamber system to grow plants till the expansion of at least 6 leaves, because the emergence of nodal roots is comparatively late in sorghum (Vijaya et al., 2010).

4.1.1   EFFECTS OF HERBIVORE ON PLANT GROWTH AND REPRODUCTION

The evolution of tolerance is one potential plant response to selection imposed by herbivores. The plant architecture and sectoriality influence plants' ability to tolerate loss of tissue. However, each may either limit or assist a plant's capability to make up for an herbivore attack based on the damaged part of the plant and the identity of the damaging herbivore. The impacts of herbivore on plant growth and reproduction would be understood by studying specific characteristics in species which demonstrate differential tolerance. Particularly, intraspecific variation in tolerance has been recorded for individuals within and among populations with different grazing histories. A number of traits associated with sectoriality and architecture might have contributed to this variation in tolerance. These traits comprise the distribution of leaves and buds, the ability to release secondary meristems from dormancy, and the timing of resource transfer both before and after damage (Robert, 1996).

4.1.2   GENE ACTION AND GENE EFFECTS CONTROLLING PLANT HEIGHT

In sorghum and maize the genetic location and effects of genomic regions regulating plant height were determined by using restriction fragment length polymorphisms (RFLPs). F_2 plants (152) obtained from the cross

CK60 x PI229828 was used for the study. Among ten linkage groups (LGs) overlaying 1299 cM, 106 genomic, and cDNA clones were identified. In each LG, four regions corresponding to previously identified *dw* loci were identified through Interval mapping. Further, these regions for plant height reported in sorghum found to be orthologous to those earlier reported in maize. Among genomic regions significant variations were observed in gene effects and gene action. About 9.2–28.7% of the phenotypic variation was observed in each region and PI229828 alleles provided increased plant height. Tallness was found to be dominant or over dominant and imparted by alleles from PI229828 for three quantitative trait loci (QTL). At the fourth QTL, though PI229828 gave improved plant height, short stature was partly dominant. A multiple model revealed that these four regions together accounted for 63.4% of the total phenotypic variation. Therefore, this study has given an insight into gene action and gene effects controlling plant height in sorghum, which could be utilized in backcross breeding aimed at germplasm conversion (Pereira and Lee, 1995).

4.1.3 DUS TESTING

DUS testing was undertaken in sorghum using 74 sorghum indigenous cultivar (IC) and one sorghum exotic cultivar (EC). These cultivars were characterized by 27 morphological descriptors given by PPV and FRA for DUS testing in sorghum. Results unveiled that there was maximum variation for glume characteristics such as color (6 groups), the neck of panicle (5 groups), length for a flower with pedicel (5 groups), time of panicle emergence (4 groups), the color of dry anther (4 groups), panicle length of branches (4 groups), panicle shape (4 groups) and caryopsis color (4 groups) among genotypes. Based on these DUS traits genotypes were classified that helped in recognition of key characteristics of various genotypes (Deepak et al., 2018).

4.1.4 SPATIAL SIMULATION MODEL TO STUDY MORPHOGENESIS

A spatial simulation model was developed to study the morphogenesis of sorghum (*Sorghum bicolor* (L.) Moench). During the development of plants, three-dimensional (3D) measurements were made to obtain the functions and parameters depicting structural development. Then, to make

3D virtual sorghum plants these functions were expressed as morphogenesis specifications in the L-system formalism and the specifications were interpreted by specialized software. At initial plant stages, the realistic images of the lengths and shapes of successive leaves were generated by using the length and height of the early leaf collar. This model captures the dynamic interaction between partitioning and morphogenesis and provides the multifaceted results as images that help in rapid interpretation (Kaitaniemi et al., 2000).

4.1.5   QTL FOR THE TARGET TRAITS

Stress trails were undertaken to study the genetic basis of drought tolerance in sorghum. Genetic basis of grain yield and stay green in sorghum were studied under well-watered and water-stressed conditions to validate QTL detected earlier. As detection of QTL for the target traits may be influenced by the variations in phenology and plant height, QTL for flowering time and plant height were taken as cofactors in QTL studies for yield and stay green. They spotted all but one of the flowering times QTL adjacent to yield and stay green QTL. Two-plant height QTL exhibited similar co-localization. QTL for yield were detected on chromosomes 2, 3, 6, 8, and 10 with the help of flowering time/plant height cofactors. However, QTL for stay-green detected on chromosomes 3, 4, 8 and 10 were not associated with variations in flowering time/plant height. The physical locations of markers in QTL regions projected on the sorghum genome suggested that the earlier identified plant height QTL, Sb-HT9-1, and Dw2, along with the maturity gene, Ma5, had a major influence on the expression of yield and stay green QTL. Co-localization between an apparently new stay-green QTL and a yield QTL on chromosome 3 suggested that there is a possibility for indirect selection on the basis of stay-green to increase drought tolerance in sorghum. This QTL study was performed with a moderate size population and covered a limited geographic range, but still the findings strongly put emphasis on the need for adjustments for phenology in QTL mapping for drought tolerance traits in sorghum (Sabadin et al., 2012).

Roots enable plants to acquire macro- and micronutrients essential for the productivity of plants and anchorage in the soil. Phosphorus (P) is hardly available for plants due to its rapid immobilization in the soil.

Adaptation to P shortage depends on variations in root morphology in the direction of rooting systems suitable for topsoil foraging. The spatial arrangement of the network including primary, lateral, and stem-derived roots is termed as Root-system architecture (RSA). This is vital for adaptation to stress conditions. Phenotyping of RSA is a difficult task and key for studying root development. Here, 19 characters depicting RSA were examined in a diversity panel containing 194 sorghum genotypes. These genotypes were fingerprinted using 90-k single-nucleotide polymorphism (SNP) array and cultivated under low and high P availability. Multivariate analysis unveiled three different types of RSA: (1) a small root system; (2) a dense and bushy rooting type; and (3) an exploratory root system, which may assist plant growth and development in the scarcity of water, nitrogen (N) or P. Though some genotypes exhibited alike rooting types in diverse environments, others responded to P scarcity positively with the development of more exploratory root systems, or negatively by suppressing root growth. Significant QTL ($P < 2.9 \times 10^{-6}$) were located on chromosomes SBI-02, SBI-03, SBI-05 and SBI-09 using genome-wide association studies. Co-localization of these QTL and their associations with other traits showed hotspots regulating root-system development on chromosomes SBI-02 and SBI-03. The study concluded that the sorghum genotypes with a dense, bushy, and shallow root system offer potential adaptation to P scarcity in the field by permitting exhaustive topsoil foraging, whereas genotypes with an exploratory root system may be beneficial under N or water stress conditions (Sebastian et al., 2018).

4.1.6   FIELD-BASED ROBOTIC PHENOTYPING SYSTEM

Sorghum (*Sorghum bicolor*) is an important raw material for biofuel production. The measurement of yield constituent traits such as plant height, stem diameter, leaf angle, leaf area, leaf number, and panicle size in the field manually is the present-day best practice to improve sorghum biomass yield via genetic research. However, this practice is laborious, time-consuming, and turns into a bottleneck reducing experiment scale and data acquisition frequency. Here, a high-throughput field-based robotic phenotyping system is presented, which performed side-view stereo imaging for dense sorghum plants with different plant heights during the growing season. This study explained the potential of stereo vision for field-based three-dimensional plant phenotyping when current advances in stereo matching algorithms were integrated. They developed a robust data processing pipeline to measure the differences or morphological traits in plant architecture which comprised plot-based plant height, plot-based plant width, convex hull volume, plant surface area, and stem diameter (semi-automated). The measurements obtained from these images were highly repeatable and strongly correlated with the manual measurements taken in field. In the meantime, the manual collection of data for the same traits needs a large amount of manpower and time when compared to the robotic system. The finding suggested that the system could be a promising tool for widespread field-based high-throughput plant phenotyping of bioenergy crops (Bao et al., 2019).

4.1.7   FIELD-BASED HIGH THROUGHPUT PLANT PHENOTYPING (HTPP)

Plant height is a key morphological and developmental character that directly indicates overall plant growth and is extensively used for the prediction of final grain yield and biomass. At present, the manual measurement of plant height is laborious and has become a major problem in genetics and breeding programs. The main objective of this study was to assess the performance of five different sensing technologies for field-based high throughput plant phenotyping (HTPP) of sorghum height. So, to measure the plant height in the field, the performance baselines were obtained by evaluating (1) an ultrasonic sensor, (2) a LIDAR-Lite v2 sensor, (3) a Kinect v2 camera, (4) an imaging array of four high-resolution cameras on a ground vehicle platform, and (5) a digital camera on an unmanned aerial vehicle platform. Data was recorded on 80 single-row plots of a US × Chinese sorghum recombinant inbred (RI) line population. Different percentiles of elevation observations within each plot were averaged to get the plot-level height. Device cost, measurement resolution, and simplicity and effectiveness of data analysis of sensing technologies were compared to evaluate the functioning of each sensing technology qualitatively. The heights measured by the ultrasonic sensor, the LIDAR-Lite v2 sensor, the Kinect v2 camera, and the imaging array showed a strong correlation with the manual measurements ($r \geq 0.90$). However, a relatively lower correlation was observed between the heights measured by remote imaging and by manual method (r = 0.73). The study findings established the ability of the suggested methods for precise and efficient HTPP of plant height and can be extended to other crops. The evaluation method deliberated in this study could assist the field-based HTPP research at large (Wang et al., 2018).

4.1.8   GENOME-WIDE ASSOCIATION STUDY (GWAS) OF SORGHUM PLANT ARCHITECTURE TRAITS

Sorghum is getting substantial attention as a lignocellulosic feedstock due to its high water-use efficiency and biomass yield potential. The improvement of genotyping and sequencing technologies made genome-wide association study (GWAS) consistently used method to study the genetic mechanisms that bring about natural phenotypic variation. This study was undertaken to determine the overall genetic architecture of grain and biomass related plant architecture characters and the given association of allelic variants in gibberellin (GA) biosynthesis and signaling genes with these phenotypes. For this GWAS was performed for nine plant architecture traits related to grain and biomass. Total 101 single-nucleotide polymorphism (SNP) characteristic regions were linked with at least one of the nine traits, and two of the important markers resembled to GA candidate genes, GA2ox5 (Sb09 g028360) and KS3 (Sb06 g028210), influencing plant height and seed number, respectively. Previously QTL for leaf angle was reported on chromosome 7 and its resolution was increased to a 1.67 Mb region comprising seven candidate genes with good potentials for additional research. New information on the relationship of GA genes with plant architecture traits and the genomic regions regulating variation in leaf angle, stem circumference, internode number, tiller number, seed number, panicle exertion, and panicle length was provided in this study. Apart from this, the GA gene influencing seed number variation (KS3, Sb06 g028210) and the association of genomic region on chromosome 7 with variation in leaf angle were significant conclusions of this study. Therefore, the study represented that the establishment of future validation studies requires the application of this information in breeding programs (Jing et al., 2016).

4.2   IDEAL PLANT TYPES FOR HIGHER PRODUCTIVITY OF SORGHUM

4.2.1   SWEET SORGHUM IDEOTYPES

Sweet sorghum has higher levels of directly fermentable reducing sugars within the culm and the ability to gather high biomass under low-input production systems when compared to other potential feedstocks such as sugarcane, sugar beet, maize, and watermelon. Further, it can be cultivated on marginal lands due to its higher tolerance to drought and the effective use of solar radiation and nitrogen-based fertilizers than maize and sugar cane. All these collectively make sweet sorghum a biofuel crop of huge potential. New phenotypes developed during plant domestication and sustained crop improvements through artificial selection make up the domestication syndrome. In this paper, Sylvester et al. (2015) draw a similarity and introduced the term biofuel syndrome to denote a set of sweet sorghum characters, such as plant architecture (root, leave, and stem), flowering time and maturity along with biomass bioconversion efficiency, that are related with the production of biofuel and to differentiate it from grain and forage sorghum characters. They discussed the biofuel syndrome open for targeted genetic change viz. the genetics and genomics of these characters as a potential way to improve sweet sorghum for biofuel production. The primary factors in the exploitation of sweet sorghum as a bioenergy crop are the continuous availability of sweet sorghum, transport, and storing abundant quantity and minimum postharvest loss of fermentable sugars. Owing to the comparatively short history of sweet sorghum breeding, they considered the development of ideotypes adaptable to different phonological needs to maximize the quick exploitation of sweet sorghum for biofuel production.

4.2.2   ROOT SYSTEM ARCHITECTURE (RSA)

The root system architecture (RSA) and prospects and limitations for the genetic improvement of sorghum crop were described. Abiotic stresses such as global climate change and shortage of water and nutrients reduces crop yield significantly. The manipulation of RSA in the direction of a distribution of roots in the soil that enhances water and nutrient uptake can minimize the negative effect of these factors on yield. It was reported that most of the genetic variation for RSA is determined by a set of QTL.

Sophiede-Dorlodot et al. (2007) discussed that marker-assisted selection (MAS) and cloning of QTL for RSA are in progress using genomic resources, candidate genes and the knowledge acquired from *Arabidopsis*, rice, and other crops. However, effective and precise phenotyping, modeling, and collaboration with breeders remain significant challenges, especially when characterizing ideal RSA for diverse crops and target environments.

4.2.3 STAY-GREEN SORGHUM HYBRIDS

It was reported that sorghum breeding programs in the USA and Australia used stay-green, i.e., maintenance of green leaf area at maturity (GLAM), as an indicator of post-anthesis drought resistance. Field trials were conducted on a cracking and self-mulching gray clay. Nine closely related hybrids with different rate of leaf senescence were cultivated under two water-stress treatments viz. post-flowering water deficit and terminal (pre- and post-flowering) water deficit, and a fully irrigated control. Under the terminal water deficit, a positive correlation was observed between grain yield and GLAM. The rate of leaf senescence had a negative correlation with grain yield. Under water-limited conditions, delayed onset of leaf senescence, i.e., 76 DAE increased grain yield by ≈0.35 Mg ha^{-1}. Stay-green hybrids gave 47% more post-anthesis biomass than their senescent counterparts (920 vs. 624 g m^{-2}) in the terminal water deficit treatment. Among eight of the nine hybrids significant differences were not found in grain yield under fully irrigated conditions. This indicated that yield is not limited by the stay-green trait in the well-watered control. The study findings revealed that sorghum hybrids with the stay-green trait have a significant yield advantage under post-anthesis drought when compared with hybrids lacking this trait (Andrew et al., 2000).

Sorghum is a major cereal crop cultivated for grain, forage, sugar, and bioenergy production. Though tall, late-flowering landraces were preferred in Africa, short early flowering varieties were grown in US grain sorghum breeding programs to decrease lodging and to enable machine harvesting. Four loci affecting stem length (Dw1–Dw4) were identified.

Further investigations unveiled that ABCB1 auxin transporter is encoded by Dw3 and a highly conserved protein that regulates cell proliferation is encoded by Dw1. In this study, Dw2 was detected using fine mapping and then confirmed by sequencing the Dw2 alleles in Dwarf Yellow Milo and Double Dwarf Yellow Milo, the progenitor genotypes where the recessive allele of dw2 originated. They found that Dw2 locus resembles Sobic.006G067700, a gene that encodes a protein kinase homologous to KIPK, a member of the AGCVIII subgroup of the AGC protein kinase family in Arabidopsis (Hilley et al., 2017).

An experiment was performed to verify the association between morphological characters that can be used as the criteria to select sorghum lines with high grain yield and earliness. 160 sorghum elite lines were evaluated for 18 traits. The estimations of phenotypic and genotypic correlations between the traits were expressed graphically using a correlation network. Two path analyses were conducted, in the first grain yield was taken as principle dependent variable, while in the second analysis flowering was principle dependent variable. On the whole, the evaluated traits explained most of the variation in the grain yield and flowering of sorghum lines. Selection of sorghum lines with a wider third leaf blade from flag leaf, panicle weight, and panicle harvest index increased grain yield. Selection of sorghum genotypes with higher plant height reduced earliness and increased grain yield. Therefore, the study findings suggested the formation of selection indices intending at increasing the grain yield and earliness in sorghum genotypes simultaneously (Da Silva et al., 2017).

Modifications in plant architecture, particularly conversion to compact canopy for cereal crops, significantly increased the grain yield in wheat (*Triticum aestivum*) and rice (*Oryza sativa*). In sorghum (*Sorghum bicolor* L. Moench) which is a multipurpose crop with an open canopy, plant architecture is a key characteristic considered for modification. Here, a sorghum genetic stock, KFS2061, a stable mutant with short and erect leaves developing compact plant architecture was characterized. Genetic analysis of an F_2 population obtained from the cross of KFS2061 to BTx623 revealed that the shortleaf is recessive and seemed to be regulated by a single gene. The shortleaf trait expression initiated with the 3rd leaf and is propagated through the entire leaf hierarchy of the canopy. Constant steep leaf angle, 43° (with the main culm as reference), and greener leaves were observed in shortleaf mutant. Biochemical analyses showed that the chlorophyll and cellulose content per leaf area is higher in the mutant than the wild type. Decrease in cell length along the longitudinal axis and expansion of bulliform cells in the adaxial surface of the mutant leaf was unveiled in histological studies. Further assessment of agronomic characters specified that this mutation could increase harvest index. Thus, the study presented evidence on a short leaf genetic stock that could provide a fundamental source to understand how to modify plant canopy architecture of sorghum (Gloria et al., 2013).

4.3   CONCLUSIONS

Ideal plant types should have desired plant features such as leaf morphology, orientation, branching patterns, canopy for effective capture of sunlight that increases the productivity of crops. Different authors proposed several studies on architecture and ideal plant types of sorghum for the improvement of grain yield. In this chapter, different research activities on sorghum plant architecture and ideal plant types were briefly summarized.

KEYWORDS

- **architecture**
- **indigenous cultivar**
- **planting density**
- **quantitative trait loci**
- **restriction fragment length polymorphisms**
- **sorghum**

REFERENCES

Andrew, K. B., Graeme, L. H., & Robert, G. H., (2000). Does maintaining green leaf area in sorghum improve yield under drought?: II. Dry matter production and yield. *Crop Science, 40*, 1037–1048.

Bao, Y., Yin, T., Lie, B., Matthew, W., Salas, F., Maria, G., Schnable, & Patrick, S., (2019). Field-based robotic phenotyping of sorghum plant architecture using stereo vision. *Journal of Field Robotics, 5*, 78–85.

Da Silva, K. J., Teodoro, P. E., De Menezes, C. B., Júlio, M. P. M., De Souza, V. F., Da Silva, M. J., Pimentel, L. D., & Borém, A., (2017). Contribution of morph agronomic traits to grain yield and earliness in grain sorghum. *Genetics and Molecular Research, 5*, 89–94.

Deepak, R. P., Pahuja, S. K., Verma, N. K., & Shalu, C., (2018). Morphological characterization of sorghum [*Sorghum bicolor* (L.) Moench] germplasm for DUS traits. *International Journal of Current Microbiology and Applied Sciences, 7*, 10–17.

Gloria, B., Cleve, F., Zhanguo, X., & John, B., (2013). Developmental and genetic characterization of a short leaf mutant of sorghum (*Sorghum bicolor* L. (Moench)). *Plant Growth Regulation, 71*, 271–280.

Hilley, Josie, L. W., Brock, D. T., Sandra, K. M. C., Ryan, F. M., Ashley, J. M. K., Brian, A. M., et al., (2017). Sorghum Dw2 encodes a protein kinase regulator of stem internode length. *Scientific Reports, 7*, 4616.

Jing, Z., Maria, B., Mantilla, P., Jieyun, H., & Maria, G. S. F., (2016). Genome-wide association study for nine plant architecture traits in sorghum. *The Plant Genome, 9*, 45–56.

Kaitaniemi, J. P., Hanan, S., & Room, P. M., (2000). Virtual sorghum: Visualization of partitioning and morphogenesis. *Computers and Electronics in Agriculture, 28*, 195–205.

Parra-Londono, S., Mareike, K., Birgit, S., Rod, S., Silke, W., & Ralf, U., (2018). Sorghum root-system classification in contrasting P environments reveals three main rooting types and root-architecture-related marker-trait associations. *Annals of Botany, 121*, 267–280.

Robert, J. M., (1996). Plant architecture, sectoriality, and plant tolerance to herbivores. *Vegetation, 127*, 85–97.

Sabadin, P. K., Malosetti, M., & Boer, M. P., (2012). Studying the genetic basis of drought tolerance in sorghum by managed stress trials and adjustments for phenological and plant height differences. *Theoretical and Applied Genetics, 124*, 1389–1402.

Sophiede, D., Brian, F., Loïc, P., Adam, P., Roberto, T., & Xavier, D., (2007). Root system architecture: Opportunities and constraints for genetic improvement of crop. *Cell, 474*–483.

Sylvester, E. A., Li-Min, Z., & Yan, X., (2015). Sweet sorghum ideotypes: Genetic improvement of stress tolerance. *Food and Energy Security, 1*, 25–34.

Vijaya, S., Erik, J. V., Oosterom, David, R. J., Carlos, D. M., Mark, C., & Graeme, L. H., (2010). Morphological and architectural development of root systems in sorghum and maize. *Plant Soil, 333*, 287–299.

Wang, X., Singh, D., Marla, S., Morris, G., & Poland, J., (2018). Field-based high-throughput phenotyping of plant height in sorghum using different sensing technologies. *Plant Methods, 14*, 53.

Yonghong, W., & Jiayang, L., (2008). *Molecular Basis of Plant Architecture*. Institute of Genetics and Developmental Biology, Chinese Academy of Sciences, Beijing – 100101, China.

CHAPTER 5

Sorghum Botany

ABSTRACT

This chapter provides a brief summary of different aspects of botany such as taxonomy, morphology, the anatomy of roots, stems, their characterization, classifications, and relation to adaptation to abiotic stresses such as drought, a flood is presented. It also discusses the molecular characterization of these aspects.

5.1 BOTANY

Berenji and Dahlberg (2004) reviewed about the status and offered a perspective on uses of sorghum [*Sorghum bicolor* (L.) Moench] use and on breeding strategies followed in Europe. He emphasized his discussion on its botany, origin, utilization, agronomic performances, and improvement. He specified that the users as well as the producers of sorghum must be better known with the crop. This familiarity with sorghum enables in taking the advantage of specific key characteristic traits of sorghum for utilization and its improvement in production as related to European climatic conditions.

5.1.1 STANDARD SET OF GROWTH STAGES

Research studies oriented towards improvement of sorghum yields or those involved in development of hybrids in grain sorghum carry out several samplings during the growth cycle. Most of these samples that are taken are generally based on the calendar date, days after planting or emergence, or plant height, though they does not have the exact relationships with that of the morphological or physiological age or the plant status. In sorghum though, some stages of its growth were reasonably well recognized, the growth cycle of sorghum has not been fully explained. Vanderlip and Reeves (1972) mentioned that there is a need to define some standard set of growth stages. Therefore, in this aspect they have conducted some studies of grain sorghum hybrids of different maturities and defined the standard ten stages of development and also showed their importance. These include emergence, three-leaf, five-leaf, growing-point, differentiation, final leaf visible in whorl, boot, half-bloom, soft dough, hard dough, and physiological maturity. Thus, they mentioned that these stages can be taken as standard stages in describing the time of sampling or treating sorghum.

Amasalu and Endashaw (2000) evaluated a total of 415 sorghum (*Sorghum bicolor* (L.) Moench) accessions collected from different regions of Ethiopia, Eritrea, and a set of introduced lines were assessed for 15 quantitative characters to ascertain the extent and geographical pattern of morphological variation. The amount of variation was highly prominent for agronomically significant characters of sorghum. These characters comprised plant height, days for 50% flowering, peduncle exertion, panicle length and width, number, and length of primary branches per panicle and thousand seed weight. Most of the characters had considerable variation. The findings suggested that environmental factors like altitude, rainfall, temperature, and growing period play a significant role in regional variation. Average plant height and days for 50% flowering showed clinal variation along the gradients of rainfall pattern and growing period in Ethiopia, further, most of the characters had considerable positive correlation coefficients. This includes the association between agronomic characters of principal importance in sorghum breeding such as plant height and days for 50% flowering and also between various characters and the altitude of the collection sites.

Sweet sorghum (*Sorghum bicolor*) is analogous to common grain sorghum with a sugar-rich stalk (Figure 5.1). Sweet sorghum is characterized by wide adaptability, drought resistance, water logging tolerance, saline-alkali tolerance, rapid growth, high sugar accumulation, and biomass. The major sources of sugar production in the world are sugarcane (*Saccharum officinarum*) and sugar beet (*Beta vulgaris*) both of them has long growth period and high water requirements. These two features of them are making them a disadvantageous source for sugar production. Further, higher prices, water, and air pollution caused by molasses have turned out the search towards the use of sweet sorghums in the sugar production. Further these sweet sorghums have less water requirements compared to sugarcane and beetroot.

FIGURE 5.1 Sorghum field.

5.1.2 *RATE OF SPONTANEOUS CROP TO WEED HYBRIDIZATION*

Weed species of sorghum have specific characteristics which are persistent and stable. In many cases, during the hybridization and wide-scale commercial release of transgenic crops or crop weed hybrids the importance of the gene flow between crops to weed is often overlooked and controversies persist on this aspect. There is also a possibility of spontaneous crop to weed hybridization to take place in many crop/weed systems and in many of the cases this is yet unknown. Availability of relevant data on the development of crop/weed hybrids will be of present-day significance when one has to consider for the wide-scale commercial release of transgenic crop plants and the probable escape of engineered genes via crop to weed hybridization. Paul et al. (1996) examined the incidence and degree of spontaneous crop to weed hybridization between *Sorghum bicolor* and *S. halepense*, Johnson grass. The hybrid plants were identified through progeny testing by using an isozyme marker. The experimental results have indicated that there is high variability in the incidence and rate of hybridization with respect to the weed distance from the crop, study site, and year of the study. They detected crop/weed hybrids at distances of 0.5–100 m from the crop. They specified that in this system the interspecific hybridization occurred at a considerable and quantifiable rate. Thus, the transgenes introduced into crop sorghum are likely to have the chance to escape cultivation through interspecific hybridization with Johnson grass. Some characteristics proved to be more beneficial to weeds holding them. These beneficial traits in weeds can be expected to persist and spread. Therefore, they suggested that while developing biosafety guidelines for the commercial release of transgenic sorghums, this issue needs to be addressed without fail.

5.1.3   PRESERVATION OF LANDRACE DIVERSITY

Availability of landraces of crops though widely exist, in many crop plants these landraces were decreased as they were not preserved by the farmers in the past. In Shewa and South Welo regions of Ethiopia the traditional farmers of those regions were thought to preserve the sorghum landrace diversity. The study undertaken by Teshome et al. (1999) has confirmed this. In their study, they found that sorghum landrace diversity at the field level had important associations with the number of selection criteria used by the farmers, field altitude, field size, pH, and clay content. With an increase in the number of selection criteria an increase in the landrace diversity was observed in the fields.

5.1.4   STRIGA RESISTANCE CHARACTERISTICS OF SORGHUM

Maiti et al. (1984) studied in ten sorghum cultivars the method of *Striga* parasitization, and also the factors involved in conferring resistance in resistant cultivars. Most of the *Striga haustona* failed to penetrate beyond the endodermis in the resistant cultivars however, these haustona in the susceptible cultivars penetrated deep into the endodermis. Further, it was found that in the resistant cultivars there was a deposition of silica in the endodermal cells which exhibited marked variations in the endodermal and pericyclic thickening among the resistant and susceptible cultivars. Extra thickening in pericyclic cells restricting the entry of haustonum into the endodermis was seen in the sorghum cultivars of N-13 and IS-4202. Ten cultivars also exhibited variations in the differential haustorial reactions. These reactions involved pericycle thickening in response to haustonal infestation, haustonal collapse, tyloses-like occlusions in the xylem vessels, and the deposition of dark-staining materials in the cortex. Though definite conclusion could not be drawn about the association between the extent of mechanical tissue development and field resistance, there was evidence that some field-resistant cultivars have strong mechanical tissues.

5.2   TAXONOMY

5.2.1   FOLK TAXONOMY

The poorest as well as the food insecure people grow sorghum as one of the main staple food crops. In the world this staple food crop was under cultivation for thousands of years in Ethiopia which was supposed to be the center of origin and diversity for sorghum crop. Thus, in this region, indigenous knowledge-based sorghum classification and naming have a long tradition where the farmers were cultivating sorghum for at least 500 years (20 generations). Firew (2007) in his study noted that sorghum is named as *Mishinga* in the region. Farmers used 25 morphological, sixty biotic and abiotic and twelve use-related traits in folk taxonomy of sorghum. Farmers categorized their gene pool by hierarchical classifications into parts that represented distinct groups of accessions. Folk taxonomy trees were made in the highland, intermediate, and lowland sorghum ecologies. More than 78 folk species have been recognized. The folk species were named after morphological, use-related, and breeding method used. He described the comparative distribution of folk species over the region, folk taxonomy stability, and comparison of folk and formal taxonomy. He identified new folk taxonomy descriptors and suggested them for use as formal taxonomy descriptors. It is specified that combined folk-formal taxonomy has to be used for better collection, characterization, and utilization of on farm genetic resources.

The indigenous component in Australian sorghums includes 17 species and 1 variety of which 14 species and the variety are endemic, and 8 taxa are new. *Sorghum brevicallosum* is reduced to a synonym of *S. timorense*, which also includes *S. australiense*. Four earlier recognized, subgenera were accepted with modified circumscriptions and floristic compositions. On morphological evidence, subgenus *Stiposorghum* stand for the most advanced members and subgenus *Para-Sorghum,* the most primitive. Some taxa were polymorphic; others show distinctive features. Characters pertaining to pubescence, pruinosity, nervation, lodicules, and caryopsis were deliberated as unspecialized characters. First chromosome counts were recorded for eight species. Indigenous species were characterized with polyploidy, which include diploids, tetraploids, hexaploids, and octaploids. Ploidy levels were found to be consistent in many of the species though they exhibited some variations in few species. The *S. macrospermum* has significantly smaller chromosomes than any other indigenous species. Cleistogamy takes place in *S. laxiflorum*. Some species were habitat-specific; many are widely adaptable. All the annual species but few have limited distributions. Seed dormancy and germination, the effects of fire, and patterns of vegetative and floral development phases were ecological aspects discussed. From nutrition point of view the herbage of both annual and perennial species is deficient in macronutrients, and lacks sufficient N and P to maintain beef cattle. Only the N and P concentrations in seed heads of *S. macrospermum* were similar to those in the grain of the cereal *S. bicolor*. These seed heads of *S. macrospermum* in the past were essential source of starchy food for Aborigines. Thus, all these indigenous taxa forma genetic resource for potential utilization by plant breeders (Lazarides et al., 1991).

Fourteen phenotypic characters were selected to obtain taxonomic proof on the similarities of 177 accessions of sorghum from North Shewa and South Welo regions of Ethiopia (Teshome et al., 1997). They conducted the canonical discriminant analysis (CDA) and MODECLUS cluster analysis to verify whether the 177 accessions could form clusters on the basis of their morphological characters, and also to test the reliability of farmers' naming of the five most common sorghum landraces represented by 44 accessions. Their Multivariate analyses grouped the 177 accessions into three clusters connected by a few phenotypic intermediate landraces. For easy classification of sorghum crop plants they established a botanical key. Many accessions of the five most common landraces named by the farmers were divided into dissimilar groups; this suggested that the naming of these sorghum landraces by farmers' is consistent. Midrib color, grain color, grain size, glume color, glume hairiness, and grain shape were the primary morphological characters utilized by the farmers to name these sorghum landraces.

Monaghan (1979) reviewed the literature on different aspects of the Johnson grass, *Sorghum halepense* (L.) Pers biology. He deliberated its distribution, cytology, taxonomy, life cycle, variability, and the dormancy and germination of its seeds and rhizomes.

Garber (1950) has made a comprehensive survey of the cytology and taxonomy of the *Sorghastrae*. It includes the subgenera *Eusorghum, Chaetosorghum, Heterosorghum, Sorghastrum, Parasorghum,* and *Stiposorghum* of the genus *Sorghum,* and the genus *Cleistachne.* He compared the chromosome morphology, numbers, and associations in the *Sorghastrae.* He has given a cytological key to the species of *Parasorghum* and *Stiposorghum.* Topics on the interspecific hybrid between *S. leiocladum,* and *S. nitidum,* within *Parasorghum* and the possible origin of *S. nitidum,* an allopolypioid species; autopolyploidy and chromosome association in *Parasorghum,* as shown in a triploid of S. *purpureo-sericeum* and in *S. leiocladum* and supernumerary chromosomes in *Parasorghum* were dealt with reference to cytological aspects. He discussed the breeding behavior with reference to self and cross-pollination, seed germination and crossability in the *Sorghastrae.* Hybridization experiments were limited to *Parasorghum* and *Stiposorghum*; the two subgenera appear to be genetically isolated. The taxonomy and distribution of the *Sorghastrae* are examined, with special reference to *Parasorghum* and *Stiposorghum.* Southeastern Asia is proposed as the probable center of origin of the *Sorghastrae.* He discussed the phylogeny of the *Sorghastrae* in relation to morphological characters viz., pedicelate spikelet, panicle branching, glume nervation, awn, and callus development, and lodicules. The evidence suggests that the newly established subgenera *Chaetosorghum* and *Helerosorghum* represented by *S. macrospermum* and *S. laxiflorum* respectively warrant equal status with the other subgenera of *Sorghum.* Although there is no evidence concerning the position of *Chaetosorghum* in the phylogenetic scheme presented, the affinity of this subgenus to *Eusorghum* is regarded as unmistakable. A second evolutionary series consists of *Heterosorghum, Sorghastrum, and Cleistachne; Parasorghum* and *Stiposorghum* probably represent an isolated line of evolution in the *Sorghastrae*, a view supported by the fact that the two subgenera have a basic chromosome number of 5, in contrast to the basic number of 10 for the other four subgenera of *Sorghum*. Finally, the basic chromosome number of the subtribe Andropogoninae is discussed and it is suggested that the ancestral stock of this subtribe possessed a basic chromosome number of 5.

5.2.2 GENETIC EROSION (GE) IN SORGHUM

The Ethiopian region has an extensive range of agro-climatic conditions. The vast resources of agrobiodiversity that are found in the country are characteristically found in this region. A chief character of these resources is the occurrence of huge genetic diversity (GD) of a variety of crop plants in the country. Of these, one of the most on farm genetically diverse crop is sorghum. Since the advent of scrupulous formal breeding after green revolution, the GD in most crops has been threatened worldwide. Firew (2008) assessed the on farm genetic erosion (GE) by employing a range of research methodologies which were mainly focused on group interviews with 360 farmers, on farm monitoring and participation with 120 farmers, key informant interviews with 60 farmers and development agents, and semi-structured interviews with 250 farmers. In addition, diversity fairs were performed with over 1200 farmers. In spite of the complexity of assessing GE, it was assessed by different methods; that is, by temporal method (comparing 1960 and 2000 collections), area method, and semi-structured interview method at individual, community, or *wereda* level and causes of varietal loss from other various outlooks. Farmers observed GE as the reduced significance of the variety as shown by lower proportion in the varietal portfolio. Reduced benefit from the varieties, drought, Khat expansion, decreased land size, and introduction of other food crops were five principal factors for varietal loss at individual farmers' level. Wealth groups and ecological regions did not have any effect on the GE, though farmers did not make simple substitution as a strategic mechanism for genetic resources management. Temporal and spatial methods were used for the quantification of GE at the regional level. There was a complementation between farmer varieties (FVs) and improved varieties

(IVs). They explained the complete process of GE mainly by three models, viz.: Bioecogeographic enhanced GE model, Farmer induced GE model and Farmer-cum-bioecogeographic GE model. As aforesaid, sorghum GE behavior is totally different from other food crops such as tetraploid wheat. The prediction in the late seventies that complete erosion of FVs by IVs by the end of the eighties, the principle of GE that competition between IVs and FVs, favors the former and results in the substitution of the latter is not applicable in the context of sorghum in Ethiopia. Therefore, preservation of the on farm GD of sorghum is a reality but GE is rhetoric.

5.3 MORPHOLOGY

Multivariate methods, containing principal component, cluster, and discriminant analyses, were used by Amsalu and Endashaw (1999) to assess the patterns of morphological variation and to group 415 sorghum accessions for 15 quantitative characters. The first five principal components explained 79% of the total variation with plant height. Among all the characters, the principal component of much importance was days to 50% flowering. All these accessions by Cluster analysis were grouped into ten clusters. A larger percentage of accessions of the same adaptation zones and accessions from regions of origin with related agro-climatic conditions were grouped together. This discriminant analysis has shown that the discrimination of accessions would be more pronounced when the analysis is based on the zone of adaptation than the regions of origin. They concluded that the morphological variation in the sorghum accessions was structured mainly by environmental factors. They discussed about these in the plant breeding and germplasm conservation programs.

In the distribution of landraces in varied ecosites of North Shewa and South Welo of Ethiopia, the adaptive potential of these was mostly due to two potential characters viz., the compactness of the panicle and its shape. This has been confirmed by Adugna et al. (2002), in their investigation of 34 sorghum landraces from 1020 individual plants. These 34 landraces were grouped mainly on the basis of variables viz., administrational zones, Woredas (smallest administrative unit), ecosite of origin and altitude. These were taken as four classifying variables of these landraces. Landraces were grouped into two or more phenotypic classes and within these classes the morphological variations were estimated for fourteen characters of qualitative traits by the usage of Shanon-Weaver diversity index (H'). There was phenotypic variation both between and within each of the four categorizing variables. The H' index ranged between 0.32 to 0.98 for all the landraces. However, the overall mean Shanon Weaver diversity index for the landraces was around 0.77 ± 0.04.

Further, a considerable variation was noted for all the fourteen qualitative characters within all the classifying variables, observed through one□ way analysis of variance (ANOVA). They grouped the landraces into five clusters on the basis of Cluster analysis which was mostly based on ordinal variables. In this cluster analysis a larger proportion of landraces which shared alike altitude classes and akin ecosites were assembled together. Grouping of these landraces into a specific cluster was mostly on the basis of potential qualitative characters viz., the compactness of the panicle, the shape of the panicle and also the juiciness character of the stalks. Much of the altitudinal and the ecological differentiation were found mainly due to the two potential qualitative characters, i.e., the panicle compactness and its shape. These two characters reflected the adaptive significance of the sorghum landraces in the regions of Ethiopia.

Shannon weaver diversity index analysis of 450 accessions of sorghum collected from different locations of Ethiopia and Eritrea these accessions exhibited vide phenotypic variations in accordance to their regions of origin and adaptation zones. Through partitioning of the variation between and within regions and those of the adaptation zones has indicated that a higher proportion of the variation was confined within the regions of origin and also within the zones of the adaptation regions. However, it was observed that this variation was significantly higher among the regions where in much of the variation was accounted for the compactness of the panicle as well as the shape of the panicle. Apart from this they found that these characters were not proportionately distributed. The compact panicles were recurrently found to be distributed more in the dry regions while in wet and humid regions there was the occurrence of loose panicles. This variation in the compactness of panicles was more profound within the regions of origin. Similarly, there was also variation in the differential distribution pattern of varied panicle types within the regions of adaptation. This has indicated that panicle compactness and shape significantly contributed to the adaptiveness of the sorghum accessions and landraces. Thus these landraces were widely distributed in different regions of Ethiopia. In association with other characters, the chi square analysis has indicated that character of seed color was also randomly associated (Amsalu and Endashaw, 2004).

Amasalu Ayan et al. (2000) undertook a study to assess the amount and pattern of genetic variation distribution in eighty sorghum germplasm accessions from different regions of Ethiopia. They used 20 oligonucleotide primers of RAPD analysis which resulted in at least 147 polymorphic bands among these eighty germplasm accessions of sorghum. Each primer resulted in at least 7.35 bands. Shannon-Weaver diversity index was utilized to estimate the genetic variation among the regions of these accessions' origin. Though there were wide-ranging variations in the genetic variation of these, an overall variation was at intermediate level with an H index of 53. Total variation revealed by partitioning index was to the extent of 77% within the regions of these accessions origin while it was only 23% among the regions of origin. Likewise, a huge portion (94%) of the entire variation was observed among the accessions within the zones of adaptation rather than among the adaptation zones, which was only 6%. There was a weak differentiation of the sorghum material. Both the regional and agroecological bases exhibited weak differentiation among the sorghum accessions. They mentioned that weak differentiation might have been due to the occurrence of very high rates of outcrossing in the cultivated sorghums. Further, these cultivated sorghum accessions might have even undergone their free natural hybridization with sorghum wild and weedy relatives or it might be also due to the movement of seed brought about by human movements from one region to another. These 80 sorghum accessions exhibited an average genetic dissimilarity of 36%; however, the dissimilarity index amongst the 15 regions of origin was only 13%. The confirmation of weak differentiation was further strengthened with the failure of grouping of the accessions by cluster analysis.

5.3.1 PHENOTYPIC VARIATIONS IN THE TISSUE CULTURE REGENERATED PLANTS

Hungtu et al. (1987) regenerated the sorghum plants on MS medium of 20 genotypes from the immature embryos of 20 sorghum genotypes. They supplemented the MS mineral salts with 2,4-D, zeatin, glycine, niacinamide, Ca-pantothenate, L-asparagine, and vitamins. On substitution of IAA for 2,4-D during the regeneration of calli, they found that there was best regeneration from those immature embryos which were obtained 9–12 days after pollination (DAP). However, two genotypes exhibited variabilities in the regeneration frequencies two cultivars C401-1 and C625 had a largest re-differentiation frequencies. They found that in these cultivars their capacity to re-differentiate was heritable. This capacity of re-differentiation appeared as a dominant trait, with the involvement of at least two gene pairs. Planting of the regenerated R_0 plants and their selfed (R_1 and R_2) progenies in the greenhouses has shown that vast phenotypic variations were confined to the R_0 plants. The transmission of these phenotypic variations to the next generations was lacking. However, the phenotypic variation transmission was absent certain traits of plant height, degree of fertility and midrib color were found to be transmitted as variants in the R_1 and R_2 generations. They attributed the variation in tallness to one dominant mutant gene while those of short stature and male sterility to the recessive mutant genes.

5.3.2   GENETIC VARIATION IN WILD SORGHUM

Wild sorghum populations also have large variations. These have their importance in genetic breeding programs as well as in the genetic conservation. Wild sorghum (*Sorghum bicolor* ssp. *verticilliflorum* (L.) Moench) accessions collected from five varied geographical regions of Ethiopia were assessed by Endashaw and Tomas (2000) for the degree and distribution of genetic variation by the use of RAPD markers for 93 individuals who were acting as representatives of 11 populations. The RAPD analysis has shown the generation of 83 polymorphic bands by 9 decamer primers. Each primer generated 8–12 bands with a mean of mean of 9 bands transversely of the 93 individuals. Though there was a high degree of polymorphic bands per each primer tested, the quantity of genetic variation was low to moderate among the populations and also the populations of the geographical locations. Likewise, a low mean genetic distance (0.08) was noted. They attributed the occurrence of low genetic variation in Ethiopian wild sorghums to their reduced population size which might have been the resultant of the habitat change of these wild sorghum populations. Partitioning of the genetic variation between and inside the population as well as among and within the regions of origin revealed that there was 75% and 88% variation in the genetic variation of these, respectively.

Further, the little level of differentiation of wild sorghum populations was confirmed through the Cluster analysis of genetic distance estimates of both on population and regional bases.

5.3.3 GENETIC VARIABILITY IN SORGHUM MOROCCAN LANDRACES

Landraces of sorghum in Morocco constitute marginalized species of sorghum exhibiting large genetic variability. Thus, these landraces of sorghum if exploited for their potentiality; they can be of much use in the programs aimed at the improvement of sorghum. Leila et al. (2007) investigated the genetic variability of these landraces and established their phylogenetic relationships using RAPD and SSR markers. A higher percentage of polymorphic fragments (98%) were from ISSR primers. The ISSR markers were more effective in revealing the amount of diversity levels. The totality of the fields was distinguishable from the classifications of the Jaccard's similarity index. Thus, their analysis has revealed the genetic structure of the Moroccon landraces of sorghum in close relation to that of the micro-geographical repartition of varied fields.

5.3.4 MORPHO AGRONOMICAL DIVERSITY

A wide morpho-agronomical diversity exists within the landraces, wild relatives, and sorghum germplasm accessions. These have a larger potentiality for improvement of the sorghum crop in the future breeding programs. However, many of these were lost due to varied reasons. Their genetic variability remained untapped. Geleta and Labuschagne (2005) have evaluated 45 germplasm accessions of sorghum cultivated in the highlands of Eastern Ethiopia. They have assessed this germplasm for ten qualitative traits and grouped and tabulated the phenotypic frequencies. The Phenotypic diversity index, H', indicated the presence of little variation between localities rather than within localities. This index had a mean of 0.71, with a range of 0.36 to 0.95 for the lines of germplasm analyzed. Their results have indicated that there is a large morpho agronomical diversity amongst this germplasm, which can be conserved as a germplasm resource to have its potential use in forthcoming sorghum improvement breeding programs.

5.3.5   GENETIC DIVERSITY (GD) IN LANDRACES OF SORGHUM

Landraces of sorghum exhibit variable patterns of GD. A clear understanding of landrace diversity is useful for deciphering of evolutionary forces under domestication. Apart from it, it also has its applicability in the preservation of genetic resources for their utilization in breeding programs. Duupa farmers in a village in Northern Cameroon differentiated 59 named sorghum taxa, representing 46 landraces. In each field, seeds were sown as a mixture of landraces (mean of 12 landraces per field), providing the potential for large gene flow. What level of GD brings about the great morphological diversity seen among landraces? Given the potential for gene flow, how well defined genetically is each landrace? To answer these questions, Adeline (2007) recorded spatial patterns of planting and farmers' perceptions of landraces, and characterized 21 landraces using SSR markers. They carried out analysis by means of distance and clustering methods. They grouped the 21 landraces that were studied into four clusters. These clusters resemble to functionally and ecologically distinct groups of landraces. Within-landrace they found that 30% of the total variation was only due to genetic variation. The mean F_{st} over landraces was 0.68. This value suggests that there is occurrence of high inbreeding within landraces. Though the diversity amongst the landraces was considerable and noteworthy ($F_{st} = 0.36$), the historical factors, variation in breeding systems, and farmers' practices played a role in affecting the patterns of genetic variation. Farmers' practices are input to the preservation, notwithstanding gene flow, of landraces with different combinations of agronomically and ecologically related traits. They should be considered in approaches of genetic resources conservation and use.

5.3.6   DYNAMICS OF GENETIC DIVERSITY (GD)

Monitoring and conservation priorities can only be assessed when there is a clear understanding on the dynamics of crop GD. Sub-Saharan Africa is the center of origin of sorghum. Many Sahel countries have been faced major human, environmental, and social changes in recent decades, which are supposed to cause GE. Sorghum is the second staple cereal in Niger, a center of diversity for this crop. Niger was affected with intermittent drought periods. Most important social changes took place in this region during these last decades. Monique et al. (2010) reported on a spatio-temporal analysis of sorghum GD that has been performed in villages (71). These villages covered the rainfall gradient and range of agro-ecological conditions of the Niger's agricultural areas. They utilized 28 microsatellite markers and applied spatial and genetic clustering methods for investigating the changes in GD over a 26-year period (1976–2003). Global genetic differentiation between the two collections was very low (F_{st} = 0.0025). Major differentiation was not observed in most of the spatial clusters, as computed by F_{st}, and exhibited steadiness or an increase in allelic richness, except for two of them found in eastern Niger. The genetic clusters recognized by Bayesian analysis did not show a significant variation between the two collections in the distribution of accessions between them or in their spatial location. These findings suggested that farmers' management has globally conserved sorghum GD in Niger.

5.4   ANATOMY

5.4.1   ROOT ANATOMY AND SECRETION OF ROOT EXUDATES

Root exudates are produced by root hair cells. Many studies have shown that the root exudate production occurs as small globules manufactured within the root hair cells in association with cell organelles particularly the endoplasmic reticulum. Mart et al. (2003) investigated the location of root exudate production in Johnsongrass (*Sorghum halepense* [L.] Pers.) and SX-17 (*Sorghum bicolor* × *Sorghum sudanese*) through transmission electron microscopy and light cryoscanning techniques. Their light micrographs studies have shown that production of root exudates is exclusively by the root hairs only. This was confirmed through the technique of scanning electron microscopy also. Their Transmission electron microscopic studies hold true for the hypothesis that most of the root exudate production occurs as the cytoplasmically dense root hair cell. The root exudate production in these root hair cells occurs in connection with smooth endoplasmic reticulum and probably with the Golgi bodies. Further, through the ultrastructure studies they found that the root exudates which were seen as petite globules of cytoplasmic exudate appeared to be accumulated particularly between the cell wall and that of the plasma membrane (PM). On deposition, these exudates coalesce and form larger globules. The coalesced large globules of root exudates are then passed through the cell wall which later is released as droplets near the tips of the root hairs.

5.4.2   MORPHOGENESIS OF CELL STRUCTURES

Anatomy of secondary morphogenesis was studied by Dustan et al. (1978) in the scutellum tissues of *Sorghum bicolor*. Immature embryo scutellar cells when plated on agar medium that contained 2, 4 D resulted in the formation of shoots, few callus and few embryo-like structures. These embryos like structures that resulted later exhibited the development of distinctive sorghum embryos. These had the completely developed scutellum, coleoptile, and coleorhiza. Light and electron microscopic anatomical studies have shown that the primary scutellar cells without any intervening phase of callus have the potentiality to form the shoots and embryo-like structures. However, it has been observed that in few cases, folding of the sexual embryo may give rise to the formation of the scutellum of the secondary embryos and is not formed *de novo*, whereas in some other cases, it was found that most of these structures are formed from the single cells only. The study could not trace out any evidence of formation of organized structures from the proliferating cells of the callus; instead they found that continuous subculturing of the unorganized callus results in any growth, and merely lead to the production of purple-black pigment. This pigment quickly turns necrotic.

5.4.3   STAY GREEN QTLS IN SORGHUM DROUGHT ADAPTATION

Water limitation is a serious phenomenon of occurrence particularly during the grain filling periods, very often resulting in decline in yields. However, there are certain sorghum plants having the ability to retain their green leaves at the grain-filling period. They are able to produce yields and are resistant to lodging under these water-limited conditions. This characteristic trait referred to as Stay green has been identified and mapped on a quite an amount of chromosome regions viz., *Stg1*, *Stg2*, *Stg3*, and *Stg4*. To elucidate the functions of these quantitative trait loci (QTL) and their positive effects on grain yields under water limitations, Borrel et al. (2014) have performed a study in Southeast Queensland, Australia on Stg near-isogenic lines of sorghum (NILs) and also on their recurrent parent. They explained the functions of Stg QTLs on grain yield which arise as a result of the variations on the temporal and spatial water use patterns. These, in turn, are the resultant of variations in the dynamics of leaf area of the plants. Their findings have shown that these four *Stg* QTLs are involved in the regulation of the canopy size. The canopy regulation may be brought about by a reduction in tillering thereby leading to an increase in the size of lower leaves, or may also restrain the size of upper leaves in the canopy. In addition, regulation of canopy size may also be because of a decline in the number of leaves produced per culm. The leaf anatomy and root growth are affected in various ways by these *Stg* QTLs. *Stg* QTLS modulates the development of the canopy in numerous pathways resulting in developmental plasticity. The diminution of canopy size connected with *Stg* QTLs results in a reduction in the water demand at the stage of pre-flowering and an increase in the availability of water at the time of grain filling. With increased water availability at this time, grain yields are increased.

5.4.4 XYLEM ANATOMY IN DROUGHT ADAPTATION

Sorghum plants exhibit different types of mechanisms of adaptation to drought. Salih et al. (1998) investigated the impacts of soil moisture stress in two sorghum cultivars, Tabat (drought susceptible) and Gadambalia (drought tolerant) at −0.02 MPa (wet) and −0.75 MPa (dry) water stress conditions. The drought susceptible variety Tabat produced roots with greater root length density (RLD), late metaxylem (LMX) vessels per nodal root, leaf area, and transpiration rate. Further, in this variety there was a higher decline in above parameters at soil moisture stress at depths >−0.2 m. However, there was a reduction in leaf area and transpiration rates, under dry conditions, the drought-tolerant variety Gadambaliad had higher water extraction efficiency all through the profile (0–0.9 m) than the susceptible variety. Further, it was observed that there was 1- to 3-cell-thick layer of sclerenchyma underneath the epidermis, even with the presence of sclerenchyma sheath around the peripheral vascular bundles in the stem of Gadambaliau; these anatomical differences were not as much prominent as that of in stems.

5.4.5   PLANT WATER POTENTIALS

Plant water potentials are a measure of plant response to water stress. Measurement of water potential in the plant reports the magnitude of resistances that occur with reference to water flow. The prevalence of the resistances of water flow, their magnitude, and location are the fundamental factors required to describe the plant water status. Though the water potentials and resistances to water flow vary, measurement of water potentials within plant is difficult because of these resistances. Wayne et al. (1980) conducted some interrelated experiments to describe the water potential of a non-transpiring leaf intact to sorghum plant and measured the water potential at the junction of root to shoot. The water potentials of an enclosed, nontranspiring leaf and a nonabsorbing root in solution, both attached to an otherwise vigorously transpiring and absorbing plant, were found to be alike. Their research findings have supported the assumption that in a covered leaf, the water potentials equilibrates at a common point where the leaves and root vascular connections share together, viz., the nodal complex of the root-shoot junction or the crown. The connection resistance between plant crown and exposed leaf lamina was computed as a difference in potential between that of a covered and exposed leaf together with that of calculated individual leaf transpiration rates. A superficial decline in the connection resistance was noted with an increase in the transpiration rate. They assumed that water potential of a covered leaf is equal to that of the root xylem at the point of water absorption in these experimental plants and thus calculated the calculated radial root resistances. The radial resistances of these short root axes were found to be largely dependent on the transpiration rate. However, it was observed that those plants which had moderate to high transpiration rates, in them the resistances in the roots were larger than the resistances that prevailed in the shoots.

5.4.6   DECLINE IN BUNDLE SHEATH THICKNESS AT ELEVATED CO_2 LEVELS

Jenny et al. (2000) assessed the influence of 350 (ambient) or 700 µL L^{-1} (elevated) levels of carbon dioxide on growth and key characteristic aspects of C_4 pathway (photosynthesis, carbon isotope discrimination, and leaf anatomy) in sorghum (*Sorghum bicolor* L. Moench). The CO_2 response of photosynthesis measured by Gas-exchange analysis has shown that both efficiencies of carboxylation and the CO_2 saturated rate of photosynthesis were at low levels in those plants grown at raised carbon dioxide levels. Further, these plants had a 49% decline in the content of phosphoenolpyruvate carboxylase of leaves (area basis), though any change in the content of Rubisco was not detected, there was a 3-fold increase in C isotope discrimination in leaves of these plants. In addition these plants exhibited a higher percent (33%) in bundle sheath leakiness than ambient plants with only 24% of bundle sheath leakiness. Any differences in Quantum yields were detected at both the levels of carbon dioxide. However, it was found that there was a lower ratio of quantum yield of CO_2 fixation to PSII efficiency, in plants grown at elevated CO_2, which seem to be more only when the leaf internal was below 50 μL L^{-1}. The research findings suggested that in sorghum the decline in the efficiency of the C_4 cycle at low CO_2, suggests increased electron transport to acceptors other than CO_2. Leaf anatomical study has shown that there is a two-fold decline in the thickness of the bundle sheath cell walls in plants which were grown at elevated CO_2. Thus their research findings have indicated that sorghum exhibits considerable acclimation to augmented CO_2 concentrations.

5.4.7   CELL WALL COMPOSITION

Cell wall compositions of epidermis, sclerenchyma, vascular bundle zone, and inner vascular bundles as well as pith parenchyma cells of sorghum were found to exhibit variations both with reference to the digestion characteristics and composition. Analysis of these free dried samples done by detergent extraction procedure. John et al. (1993) has shown that the *in-vitro* dry matter (DM) digestibility (g kg^{-1} after 48 h) of cell fractions was in the order of pith parenchyma (849–906) > inner vascular bundles (794–816) > sclerenchyma (692–701) > vascular bundle zone (641–679) > epidermis (608–628). The digestibility of pith parenchyma was highest and least was observed with the cells of the epidermis. However, there was an inverse order in the total cell wall content (CWC), indigestible CWC, and lignin content among the above cell fraction extracts. Lignin concentration after 96 h estimated on a DM or cell wall basis was found to be highly associated with indigestible wall residue. Similarly, a higher proportion of 61.8 to 68.2% was found for the proportion of cell wall digested after 96 h for sclerenchyma and vascular bundle zone cells rather than the cells of the pith, even though the former has a three to five folds of higher lignin content than the lignin content of pith cells.

5.4.8   SALINITY EFFECT ON XYLEM STRUCTURE

Salinity in sorghum has an effect on the growth and water conductance. Stuart et al. (2000) in their research studies on the effect of salinity on xylem structure and water use have found that there were narrower protoxylem and metaxylem cells, with a concomitant decline in the leaf width and cross-sectional area of leaves affected by salt treatment, though there was very little or no effect on the protoxylem area per area of leaf cross-section. Further, the dye uptake studies revealed that 50% of the veins are only functional in water transport in these plants, with a large diminishing rate in the volumetric water flow. In late developmental stages, the decreased flow rates of water were observed per unit leaf mass or leaf area mostly due to a decrease in the leaf surface area. Though there was a decrease in the flow rates, measurement of leaf conductance by a diffusion porometer has shown that these flow rates were not related to the diameters of elements of either the protoxylem or the metaxylem. In contrast earlier published researches have indicated that deposition rates of water are more associated with the size of the elements of the protoxylem and flow rates in the growing leaf tissues were related to the square of the radius of the protoxylem.

5.4.9   ROOT TO SHOOT DEVELOPMENT AT DIFFERENT GROWTH STAGES

In grain sorghum [*Sorghum bicolor* (L.) Moench] the relationship that exists between root to top development at more than one stage of growth under irrigated and non-irrigated conditions was investigated by Kaigama et al. (1976). The results indicate that a key characteristic feature of root development is root penetration into the soil. In sorghum, during the early parts of the growing season there was rapid root penetration, where some traces of roots were detected at depths of 140 to 150 cm at 6 weeks after emergence and a maximum amount of 17.1 g of root DM was recorded per 1.08 dm² of soil surface in irrigated sorghum at 9–10 weeks after emergence while it was only 12.3 g in non-irrigated sorghum. There was a maximum accumulation of root DM of 1/3-row at 8 weeks after emergence. Root development varied with the irrigation treatments, more quantity of roots in the top 15 cm of the soil were seen in irrigated sorghum, however, in non-irrigated sorghum there was a larger proportion of overall root DM accumulation at a deeper depth of soil rather than at the top layers. Though there was only a minimal increase in the leaf DM after 7th week of emergence, there was a drastic increase in the stem DM which had almost reached a peak by 8 weeks after emergence, while there was a decline in the stem DM during panicle development.

Safety zone formation at the junction of root to shoot enables in offering protection to the root vessels from the embolisms that originate in the shoots, Aloni, and Griff (1991) in root anatomical studies of cereals found that cereal roots exhibit a safe root vessel system and an unsafe root vessel hydraulic root architecture, which varies with cereal species. A high degree of vascular segmentation is found in maize enabling in the formation of safe vascular zones, in sorghum and oats there is a typical development of only a primary seminal root. This primary seminal root contains unsafe vessels. These vessels are continuous through the mesocotyl and all through the first node. They mentioned that in adventitious roots, vascular segmentation is not related to overall root morphology.

5.5   CONCLUSIONS

Many researchers have been assumed on different aspects of botany such as taxonomy, morphology, anatomy of root and stem, which were associated to adaptation for abiotic stresses such as drought, flooding, etc. These research advances are briefly presented in this chapter.

KEYWORDS

- anatomy
- botany
- canonical discriminant analysis
- molecular characterization
- sorghum
- taxonomy

REFERENCES

Adeline, B., Monique, D., Eric, G., Doyle, M. K., & Hélène, I. J., (2007). Local genetic diversity of sorghum in a village in northern Cameroon: Structure and dynamics of landraces. *Theoretical and Applied Genetics, 114,* 237–248.

Adugna, A., Endashaw, B., & Awgechew, T., (2002). Patterns of morphological variation of sorghum (*Sorghum bicolor* (L.) Moench) landraces in qualitative characters in North Shewa and South Welo, Ethiopia. *Hereditas, 137,* 161–172.

Aloni, R., & Griffith, M., (1991). Functional xylem anatomy in root to shoot junctions of six cereal species. *Planta, 184,* 123–129.

Amsalu, A., & Endashaw, B., (1999). Multivariate analysis of morphological variation in sorghum (*Sorghum bicolor* (L.) Moench) germplasm from Ethiopia and Eritrea. *Genetic Resources and Crop Evolution, 46,* 273–284.

Amsalu, A., & Endashaw, B., (2004). Geographical patterns of morphological variation in sorghum (*Sorghum bicolor* (L.) Moench) germplasm from Ethiopia and Eritrea: Qualitative characters. *Hereditas, 129*(3), 195–205.

Amsalu, A., Tomas, B., & Endashaw, B., (2000). Genetic variation of Ethiopian and Eritrean sorghum (*Sorghum bicolor* (L.) Moench) germplasm assessed by random amplified polymorphic DNA (RAPD). *Genetic Resources and Crop Evolution, 47*(5), 471–482.

Andrew, K., Borell, J. E., Mullet, B., George, J. E. J., Van, O. G. L., Hammer, P. E., Klein, D. R., & Jordan, (2014). Drought adaptation of stay-green sorghum is associated with canopy development, leaf anatomy, root growth, and water uptake. *Journal of Experimental Botany, 65,* 6251–6263.

Berenji, J., & Dahlbeg, J., (2004). Perspectives of sorghum in Europe. *Journal Agronomy and Crop Science, 190,* 332–338.

Borell, A. K., Mullet, J. E., George-Jaeggli, B., Van Oosterom, E. J., Hammer, G. L., Klein, P. E., & Jordan, D. R., (2014). Drought adaptation of stay-green sorghum is associated with canopy development, leaf anatomy, root growth, and water uptake. *Journal of Experimental Botany, 65,* 6251–6263.

Dunstan, D. T., Short, K. C., & Thomas, C. E., (*1978*). The anatomy of secondary morphogenesis in cultured scutellum tissues of *Sorghum bicolor. Protoplasma, 97,* 251–260.

Endashaw, B., & Tomas, B., (2000). Genetic variation in wild sorghum (*Sorghum bicolor* Ssp. *Verticilliflorum* (L.) Moench) germplasm from Ethiopia assessed by random amplified polymorphic DNA (RAPD). *Hereditas, 132,* 249–254.

Firew, M., (2007). Infra-specific folk taxonomy in sorghum (*Sorghum bicolor* (L.) Moench) in Ethiopia: Folk nomenclature, classification, and criteria. *Journal of Ethnobiology and Ethnomedicine, 3,* 38–48.

Firew, M., (2008). Genetic erosion of sorghum (*Sorghum bicolor* (L.) Moench) in the center of diversity, Ethiopia. *Genetic Resources and Crop Evolution, 55,* 351–364.

Garber, E. D., (1950). Cytotaxonomic studies in the genus sorghum. *University of California Publications in Botany, 23,* 283–362.

Geleta, N., & Labuschagne, M. T., (2005). Qualitative traits variation in Sorghum (*Sorghum bicolor* (L). Moench) germplasm from eastern highlands of Ethiopia. *Biodiversity and Conservation, 14,* 3055–3064.

Hungtu, M., Minghong, G., & Liang, H., (1987). Plant regeneration from cultured immature embryos of *Sorghum bicolor* (L.) Moench. *Theoretical and Applied Genetics, 73,* 389–394.

Jenny, R., Watling, M. C., Press, W., & Paul, Q., (2000). Elevated CO_2 induces biochemical and ultrastructural changes in leaves of the C4 cereal sorghum. *Plant Physiology, 123*(3), 1143–1152.

John, R., Wilson, D. R., Merten, R. D., & Harfield, (1993). Isolates of cell types from sorghum stems: Digestion, cell wall and anatomical characteristics. *Journal of Science, Food Agriculture, 3,* 407–417.

Kaigama, B. K., Teare, I. D., Stone, L. R., & Powers, W. L., (1976). Root and top growth of irrigated and non-irrigated grain sorghum. *Crop Science, 17*(4), 555–559.

Lazarides, M., Hacker, J. B., & Andrew, M. H., (1991). Taxonomy, cytology, and ecology of indigenous Australian sorghums (*Sorghum* Moench: Andropogoneae: Poaceae). *Australian Systematic Botany, 4,* 591–635.

Leila, M., Mohammed, A., Ouafae, B., Driss, M., & Filali- Maltouf, A., (2007). Evaluation of genetic variability of sorghum (*Sorghum bicolor* L. Moench) in northwestern Morocco by ISSR and RAPD markers. *Comptes Rendus Biologies, 330*(11), 789–797.

Maiti, R. K., Ramaiah, K. V., Bisen, S. S., & Chidley, (1984). Comparative study of the Haustorial development of *Striga asiatica* (L.) Kuntze on sorghum cultivars. *Annals of Botany, 54,* 447–457.

Mart, A. C., Rex, N. P., Leslie, A. W., & Stephen, O. D., (2003). Anatomy of sorgoleone secreting root hairs of *Sorghum* species. *International Journal of Plant Sciences, 164*(6), 861–866.

Monaghan, N., (1979). The biology of Johnson grass (*Sorghum halepense*). *Weed Research, 19,* 261–267.

Monique, D. F., & Sagnard, J. C., (2010). Spatio-temporal dynamics of genetic diversity in *Sorghum bicolor* in Niger. *Theoretical and Applied Genetics, 120,* 1301–1313.

Paul, E. A., & Norman, C. E., (1996). Crop to weed gene flow in the genus *Sorghum* (Poaceae): Spontaneous interspecific hybridization between Johnson grass, *Sorghum halepense*, and crop sorghum, *S. bicolor. American Journal of Botany, 83*(9), 27–31.

Salih, A. A., Ali, I. A., Lux, A., Luxová, M., Cohen, Y., Sugimoto, Y., & Inanaga, S., (1998). Rooting, water uptake, and xylem structure adaptation to drought of two sorghum cultivars. *Crop Science, 39*, 168–173.

Stuart, F. B., Phong, N. T., & Wendy, K., (2000). Effects of salinity on xylem structure and water use in growing leaves of sorghum. *New Phytologist, 146,* 119–127.

Teshome, A., Baum, B. R., Fahrig, L., Torranc, J. K., Arnason, & Lambert, J. D., (1997). Sorghum [*Sorghum bicolor* (L.) Moench] landrace variation and classification in North Shewa and South Welo, Ethiopia. *Euphytica, 97,* 255–263.

Teshome, A., Fahrig, L., Torrance, J. K., Lambert, J. D., Arnason, T. J., & Baum, B. R., (1999). Maintenance of sorghum (*Sorghum bicolor*, poaceae) landrace diversity by farmers' selection in Ethiopia. *Economic Botany, 53*, 79–88.

Vanderlip, R. L., & Reeves, H. E., (1972). Growth stages of sorghum [*Sorghum bicolor*, (L.) Moench]. *Agronomy Journal, 64,* 196–200.

Wyne, S., Meyer, J., & Ritchi, T., (1980). Resistance to water flow in the sorghum plant. *Plant Physiology, 65,* 33–39.

CHAPTER 6

Physiological Basis of Crop Growth and Productivity

ABSTRACT

This chapter discusses significant research advances made on the physiological basis of sorghum growth and productivity till 2019. Besides, it deliberates different factors affecting sorghum growth and productivity starting from germination and seedling establishment, vegetative growth, and flowering, fruiting, and yields.

6.1 GROWTH AND DEVELOPMENT

Grain sorghum has a determinate growth and produces a predetermined number of leaves. It has a C4 type of photosynthesis. It has high adaptability to water-limited conditions because of its efficient carbon fixation pathway. It is usually grown as annuals; however, it is said to possess a perennial habit because of its ability to continuous tiller production.

Sorghum crop shows different growth stages. The seedling emergence and seedling growth are affected by several factors like the presence of soil crust, soil temperature, depth of planting. Methods have been developed to screen sorghum genotypes for emergence from high soil temperature and soil crust. Maiti (1980) published a book on Sorghum Science making a review of research on different aspects of sorghum from emergence to grain maturity and other aspects.

Some growth stages of sorghum are: Growth stage 1 (Commences from emergence and continues up to panicle initiation); Growth stage 2 (Begins with the emergence of panicle and extends up to panicle development); and Growth stage 3 (Includes the stage of grain filling to grain maturity). Sorghum seedling emergence takes 5 to 10 days from the time of planting to emergence. This time period varies with the variety, season, and is also dependent on the growing conditions, soil temperature and moisture; the depth of sowing and to some extent, seed vigor. Delayed emergence of the seedlings often leads to sorghum stands which are uneven with more number of skimpy plants. Such sorghum stands do not result in good yields (Gerik et al., 2003).

Kraig et al. (2019) in his book Sorghum: State of the Art and Future Perspectives stated that it is important to have complete understanding of the growth and development of sorghum to unveil its responses to environmental stress, to make valid production and management decisions. He described about Sorghum development into three major stages viz., GS-1, GS-2, and GS3, where these correspond to vegetative, reproductive, and grain fill stages, respectively. The sorghum plant spends at least a third of its life cycle period in each of these stages. It is said that the vegetative GS-1 stage in sorghum is further subdivided into three stages viz., S0, S1, and S2 which corresponds to stage of emergence, the third leaf collar, and fifth leaf collar respectively. At the growing point of differentiation of the terminal meristem instead of giving rise to additional leaves, leads to the development of panicle structures marking the beginning of the reproductive growth. This stage of differentiation of terminal meristem to panicle structures corresponds to S3 stage. Analogous to the subdivision of vegetative stage, the phase of reproductive growth is also subdivided into two stages viz., S4 and S5 corresponding to appearance of flag leaf and boot stages respectively. In sorghum, the reproductive growth transitions to grain filling stage at half bloom only, representing stage S6. At this stage of S5, it attains a maximum value of plant height and leaf area and also accumulates more than half of the dry matter (DM) in the above-ground parts. The final stage S3, i.e., stage of grain fill is further subdivided into S7 and S8 each corresponding to soft dough and hard dough stages respectively. Finally, the grain fill stage comes to an end with S9 stage; the physiological maturity, representing the cessation of further accumulation of dries matter. The S9 stage, physiological maturity is easily identifiable with a visual indicator, where there is the appearance of an abscission layer seen as a black layer quiet opposite to the embryo. Though the timing of development and transition to respective stages during life cycle of sorghum is reliant on the genetic and environmental conditions, in a medium maturing commercial hybrid growing in Central Great Plains, this time period may be approximately at an interval of 10 days to progress to next stage of development. Many research findings have shown that sorghum plant has an immense capacity to get it adapted to the prevailing environments by the production of tillers, or by adjusting its number of potential kernels in its ear head, seed setting percentage and varying the size of kernels. Owing to its vast adaptive capacity, the yields of sorghum tend to be stable relatively across environments, the different management systems and across the years.

Vanderlip and Reeves (1972) mentioned that some standard set of growth stages need to be defined. Therefore, in this view they have performed some studies of grain sorghum hybrids of different maturities and defined the standard ten stages of development and also explained their importance.

Miller (1980) in a research paper presented in the conference of FAO on Plant Production and Protection Division described about the range of variations that are seen in *Sorghum bicolor*, in various botanical characters, viz., morphology, physiology, grain size, seedling emergence, growth period, height, grain color and juice characteristics. The survey aspects of these along with other details of the system used to develop hybrids, exploiting cytoplasmic male sterility (CMS) were given.

6.1.1   ENZYMATIC BREAKDOWN OF SORGHUM ENDOSPERM

Endosperm of cereals acts as a nutritive reserve for the developing embryo at the time of seed germination and the products released after enzymatic hydrolysis of the endosperm are utilized by the developing embryo to emerge as a seedling (Figure 6.1).

Aisien (1982) investigated on the enzymic activity of sorghum endosperm during seedling growth. Sorghum grain suitable for malting had increased enzymic activities of α-amylase, endo-β-glucanase, limit dextrinase and endoprotease. In sorghum endosperm α-amylase was found as the important starch-degrading enzyme. Sorghum endosperm had higher activities of enzymes endo-β-glucanase, limit dextrinase and endoprotease rather than in the embryo at the time of seedling growth, though there was low activity of Endo-β-glucanase activity at the time of seedling growth, this might be partially accountable for the reduced degradation of the cell walls in the endosperm of the malted grain.

Aisen et al. (1983) studied the enzymatic changes during the seed germination and seedling emergence in sorghum. It has been observed that the enzyme β-glucanase was inactive to barley β-glucan. The levels of α-amylase were not regulated by gibberellic acid (GA) and the synthesis of this enzyme in sorghum occurred in the embryo. Later this enzyme exhibited its hydrolytic properties on the starch granules of the endosperm. The enzyme Limit dextrinase appears as a zymogen in the endosperm of sorghum. They also found that the amino acid releasing proteases during germination developed in the embryo and were absent in the endosperm. Likewise, the endoproteases though were detectable in the embryo their content was higher in the endosperm rather than the embryo. The study findings have specified that in sorghum during the early stages of germination the amylolytic breakdown of endosperm is very extensive in sorghum.

6.1.2   INHIBITION OF SEED GERMINATION

James et al. (1977) gave an account about the inhibitory effects of p-coumaric and ferulic acids on germination and growth of grain sorghum. These chemicals exhibited synergistic effects. The equimolar mixtures of both these resulted in a large decline in sorghum seed germination, shoot elongation, and total seedling growth than either phytotoxin caused when alone. Results obtained from many experiments which were repeated has shown a decline in the seed germination by 34% after 24 hrs while it was 59% after 48 hrs in these mixtures which contained 5×10^{-3} M p-coumaric and 5×10^{-3} M ferulic acids. However, there was not much inhibition of the seed germination in the seeds which were either phenol treated or untreated. The germination percentage was around 89–92%. The phytotoxic action of the combination approximated the inhibitory effect on germination of 10^{-2} M ferulic acid. Ferulic acid resulted in a larger decline in the seed germination than 10^{-2} M p-coumaric acid. In comparison to seed germination and seedling growth, the seedling growth was to a large extent sensitive than the seed germination. The equimolar mixture of 2.5×10^{-4} M p-coumaric and 2.5×10^{-4} M ferulic acids resulted in a larger decline in seedling dry weight. Further dilutions have shown a 1.25×10^{-4} M concentration of either phenol was stimulatory to seedling growth; however, a combination of these two produced inhibition.

Germination in osmotica can predict only the comparative ability of a cultivar to germinate in soil of low water content. Darry et al. (1980) in their study on NK 300 and M35 sorghum cultivars in different osmotica, found that there was a rapid and higher percentage of germination in NK 300 and it was also lead inhibited in its germination by ABA. It was further noted that when there was osmotic stress condition. GA brings about an increase in the germination of cultivar M35 only. It did not exhibit any increase in germination in NK 300 cultivar. Thus, they stated that inhibition in germination of seeds under osmotic stress created by D-mannitol or polyethylene glycols is not attributed only to a decrease in the uptake of water. It appeared that the decline is germination of seeds was due to an in the metabolic processes of cells even though they had ample of turgor potential needed for expansion of cells.

FIGURE 6.1   Sorghum seedling establishment.

In this paper the toxicity of binary drug mixtures and individual drugs on the germination of *Sorghum bicolor* seeds was assessed under different pH conditions with the additions of various inorganic ions. The predicted phytotoxicity values were estimated through concentration addition (CA) and independent action (IA) methods, to evaluate whether the given mixtures of drug were more phytotoxic than the individual compounds. Then the deviation from the predicted values was determined using MDR (model deviation factor). Binary mixtures of chloramphenicol with ketoprofen, diclofenac sodium, and oxytetracycline hydrochloride found to be synergistic. The pH fluctuations showed a high effect on the phytotoxicity of the diclofenac sodium and ketoprofen solutions and increased their toxicity toward *S. bicolor*. The co-presence of inorganic ions had an impact on ketoprofen, chloramphenicol, and oxytetracycline hydrochloride. It was found that most of the interactions between sorghum plants and pharmaceuticals with additional ions were antagonistic in nature; especially those computed using the IA model. However, there were few cases of overestimation viz. one case for ketoprofen and chloramphenicol, two cases for oxytetracycline hydrochloride, and four cases for diclofenac sodium (Wieczerzak et al., 2018).

6.1.3 MANAGEMENT OF SORGHUM GROWTH AND DEVELOPMENT

Gibberellins in many plants were found to induce elongation in stems, of dwarf mutants. Page et al. (1977) undertook a study in few sorghum genotypes with varying dwarf genes to understand the involvement of the external application of gibberellins on their growth and development. Application of gibberellin to whorls of vegetative plants grown on Norwood fine sandy loam has shown that there was promotion of growth of the seedlings though higher levels of application were required to bring about required final stem height. At gibberellins concentrations up to 10^{-3} M GA_3, there was a greater increase in stem height of milo and kafir group members. The increase in stem height by gibberellins in Redland varieties was only slight while its effect was not observed in hegaris group members. These results indicate that plants exhibit variability in their responses to external gibberellins applications. A drastically reduction in tillering was noticed with the continuous appli-cation of gibberellins over a period of several weeks, simultaneously there was adventitious root development. However, it was observed that when the application of gibberellins is terminated early, there was a successive development of more tillers and tillering. The results indicated that the gibberellic effects are separable on tillering and stem height. Gibberellin application at lower rates of concentration led to a reduction in tillering without any promotion of increase in stem height, like wise higher concentrations of gibberellins applications modified both these processes. Gibberellin applications though modified the tillering and stem height they did not bring about a shift in the anthesis date in any of the genotypes studied.

In recent years unmanned aerial vehicles and systems (UAV or UAS) have become more popular for agricultural research applications. UAS can acquire images with high spatial and temporal resolutions that are perfect for uses in agriculture. In this study the performance of a UAS-based remote sensing system was evaluated for quantification of sorghum crop growth parameters like leaf area index (LAI), fractional vegetation cover (FVC) and yield. A fixed-wing UAS equipped with a multispec-tral sensor was utilized to collect image data during the 2016 growing season (April–October). Flight missions were effectively conducted at 50 days after planting (DAP; 25 May), 66 DAP (10 June) and 74 DAP (18 June). These flight missions gave image data covering the middle growth period of sorghum (Figure 6.2) with a spatial resolution of about 6.5 cm. Field measurements of LAI and FVC were also collected. Using the UAS images four vegetation indices were computed. Among those indices, the normalized difference vegetation index (NDVI) exhibited a strong correlation with LAI, FVC, and yield. Empirical associations between NDVI and LAI and between NDVI and FVC were validated and found precise for assessing LAI and FVC from UAS-derived NDVI values. NDVI ascertained from UAS imagery obtained during the flow-ering stage (74 DAP) showed high correlation with final grain yield. The strong correlations between NDVI derived from UAS and the crop growth parameters (FVC, LAI, and grain yield) suggested the suitability of UAS for within-season data collection of agricultural crops such as sorghum (Shafian et al., 2018).

FIGURE 6.2   Sorghum crop at vegetative stage.

6.2   SORGHUM FLOWERING, GRAIN FILLING AND MATURITY

6.2.1   FLOWERING

The success of modern agriculture relies on the optimal flowering time (Figure 6.3). It is well known that sorghum is a short-day (SD) tropical species. It shows considerable photoperiod sensitivity and late flowering in long days. Genotypes with less photoperiod sensitivity facilitated utilization of sorghum as a grain crop in temperate zones worldwide. Rebecca et al. (2011), in their study, identified Ma_1, the major repressor of sorghum flowering in long days, as the pseudo response regulator protein 37 (PRR37) via positional cloning and analysis of *SbPRR37* alleles that reduces flowering time in grain and energy sorghum. Numerous allelic variants of *SbPRR37* were found in early flowering grain sorghum germplasm that have unique loss-of-function mutations. They have shown that in long days *SbPRR37* activated the expression of the floral inhibitor *CONSTANS* and repressed the expression of the floral activators *Early Heading Date 1*, *Flowering Locus T*, *Zea mays Centroradialis 8*, and floral induction. It was found that the expression of *SbPRR37* is light dependent and was under the regulation of a circadian clock, with peaks of RNA abundance in the morning and evening in long days. In short days, the evening-phase expression of *SbPRR37* did not occur due to darkness and it permitted the sorghum to flower in this photoperiod. Their research provided a vision into an external coincidence mechanism of photoperiodic regulation of flowering time mediated by PRR37 in the SD grass sorghum and recognized the major alleles of *SbPRR37* that are essential for the exploitation of this tropical grass in temperate zone grain and bioenergy production.

6.2.2   GIBBERELLINS IN FLORAL INITIATION

Emily and Page (1979) reported on the effects of gibberellins and far-red light on floral initiation in sorghum genotypes with varying maturity periods. The 12-h dark period alone without either the application of GA_3 or FR in the early maturing genotype of sorghum has induced flowering. However, there was hastening in the floral initiation by FR light in early genotype of sorghum, while in those two intermediate flowering genotypes, there was hastening of floral initiation by GA_3. A combination of GA_3 and FR had a well-built synergistic effect, in speeding up of floral initiation by 30 to more than 80 d in the early and intermediate genotypes. Red light (R) did not hasten flowering; FR preceded by R provided the same effect as FR alone. GA_3 stimulated stem elongation uniformly in all the maturing genotypes, irrespective of the occurrence or absence of floral initiation. It appeared that GA_3 effect on stem elongation was not dependent on floral initiation. The capacity of GA_3 to induce flowering in sorghum, a SD plant, appears to be improved by phytochrome.

FIGURE 6.3   Sorghum crop at anthesis stage (flowering).

6.2.3   PRE-HARVEST SPROUTING (PHS) AND ABA CONTENT

Pre-harvest sprouting (PHS) in crops is major problem that affect crop yields. It leads not only to losses in seed viability but also result in marked declines in grain weights. Haydee et al. (1995) aimed at having a vision into the physiological basis of PHS resistance in sorghum. This was assessed by taking into consideration the assessment of germinability, ABA content in seed embryonic tissues and also the sensitivity of the sorghum embryos to ABA during seed development in three varieties, with divergent sprouting behavior: Redland B2 (very susceptible), SC 650 (moderately resistant) and IS 9530 (very resistant). Redland B2 caryopses were able to germinate with high germination indices from early stages of development, while caryopses of IS 9530 did not present germination indices different from 0 till near physiological maturity. An intermediate pattern of behavior of above two was seen in SC 650. However, the isolated embryos of all the three varieties germinated with maximum germination indices from as early as 15 d after pollination (DAP), the variations in grain dormancy level were not paralleled by a constantly different endogenous ABA content during maturation. But, when ABA embryonic content of 35 DAP caryopses was measured in incubation, ABA level in B2 embryos after 24 h of incubation was found to be less than half that noted in IS 9530 embryo after the same period of incubation. Furthermore, B2 embryos were found to be 10-fold less sensitive to the inhibitory effect of ABA than embryos from the other two varieties. These findings explained a significant extent of variations in germinability between sprouting resistant and susceptible varieties.

The susceptibility of grain sorghum to PHS is linked to the presence of low levels of dormancy in the seeds and also to a decreased sensitivity of embryos to abscisic acid (ABA) in its inhibition of germination. Intra-specific variability for PHS may include differential regulation of ABA signaling genes. Maria et al. (2009) examined the Sorghum genes encoding homologs for ABA signaling constituents from other species (*ABI5*, *ABI4*, *VP1*, *ABI1,* and *PKABA1*) at the transcriptional and protein level (ABI5) during grain imbibitions for two sorghum lines with complementary sprouting phenotypes and in response to hormones. They observed that theABI5 had much higher transcript levels of these genes and protein levels in imbibed immature caryopses. In both genotypes, lower transcript levels of these genes and lower ABI5 protein levels were found to be responsible for dormancy loss. Exogenous ABA inhibited germination of isolated embryos but could not avert ABI5 rapid reduction confirming a role of the seed coat in regulating ABI5 levels. A number of genes involved in ABA signaling were regulated in a different way in imbibed caryopses from two sorghum lines with different PHS response before-but not after-physiological maturity. They deliberated about the role of ABI5 in the expression of dormancy during grain development.

In maturing seed of sorghum, pre-harvest sprouting is a foremost constraint (Figure 6.4). Benech et al. (1995) studied the effect of potassium nutrition on germinability, ABA content and embryo ABA sensitivity in its developing sorghum seeds. In the plants supplied with one third that of control potassium improved germinability of developing seeds was noted during the late stages of maturation (31 and 36 days after pollination (DAP)). These variations in seed germinability were also paralleled by a decreased ABA content in the seed. The sensitivity of embryo to ABA was also lower in these seeds, while there was a peak in ABA content in the intact grains at 25 DAP which later exhibited a sharp decline. This decline in the ABA content was 30–50% lesser than that in the controls at 31 and 36 DAP. However, it was observed that there was a peak in ABA content later in the developmental period (at 31 DAP) in grains of control plants. By contrast, in the grains of those which were supplied with one-ninth of the potassium supply as that of controls, the ABA content remained constant during most period of development (20–36 DAP). The sensitivity to ABA was distinctly decreased in the embryos from both the low potassium treatments. It was more evident when these embryos were evaluated at ABA concentration of 5 μM. The data showed that potassium supply affected the germinability, ABA concentrations, and embryo ABA sensitivity in developing seeds of *Sorghum bicolor*.

FIGURE 6.4   Sorghum crop at grain filling stage.

Research findings of Eduardo et al. (1997) have presented that the differences that occur between varieties with reference to pre-harvest sprouting, in their abilities to germinate on the mother plant are not just limited to the susceptibility nature for germination, but are much related to the variations in the activity of α-amylase. Further, they suggested that both GAs and ABA participate in regulating α-amylase activity in developing *Sorghum* caryopses. They examined α-amylase activity in developing caryopses from *Sorghum* varieties Redland B2 (highly susceptible to preharvest sprouting) and in IS 9530 (very resistant to preharvest sprouting). Unsprouted caryopses of Redland B2 exhibited α-amylase activity from 31 days after pollination (DAP) onwards only and the rate of activity was twenty folds higher than the resistant variety IS 9530. Incubation of the caryopses led to an increase in the activity of alpha amylase in the embryos. Between these two varieties it was observed that addition of GA$_3$ increased the sensitivities of the embryos of IS 9530 embryos rather than those of Redland B2. Similarly, treatment of the panicles with an inhibitor of gibberellin synthesis, i.e., paclobutrazol at 3 DAP resulted in a drop in the α-amylase activity of Redland B2 caryopses down to the levels that were found to exist in control caryopses of IS 9530. Contrariwise, it was found that similar treatment of the panicles with that of an inhibitor of ABA biosynthesis, i.e., fluridone resulted in an increase in the activity of α-amylase in the caryopses of both the susceptible as well as the resistant variety.

In addition to earlier studies conducted by the research findings of Nicolas et al. (2007) has shown that a disturbance in the ABA signal transduction pathway in Redland B2 bring about the low ABA sensitivity shown by embryos from this line. This was confirmed when they worked out to explain the nature of the distinct ABA sensitivities exhibited by developing embryos from sorghum [*Sorghum bicolor* (L.) Moench] lines with dissimilar PHS behavior (Redland B2, susceptible; IS 9530, resistant). They investigated two diverse assumptions for a possible mechanism: (1) a unusual functionality of the ABA signaling pathway, and (2) a diverse rate of ABA degradation/conjugation in the apoplast of embryos from these genotypes. They assessed the first probability, by using an ABA-responsive gene (*Rab17*) as a reporter of variations in endogenous ABA content. These changes were artificially induced in embryos from both genotypes through fluridone application immediately after anthesis, to decrease ABA content, and embryo incubation in the presence of ABA.

They mentioned that a defect in ABA signaling ought to be seen as a level of *Rab17* expression that is free of endogenous ABA content. Similarly, for testing the 2nd possibility at two stages of development, they isolated the embryos from both these lines and incubated them in water for varying periods. Quantification of ABA concentrations in embryos and in the incubation media by radioimmunoassay has shown that, *Rab17* expression for the resistant IS 9530 line did not respond to variations in ABA levels. The ABA degradation/conjugation rates in embryos and incubation media had no significant differences between sorghum lines for any of the developmental stages examined.

During the development and seed maturation stages, if the developing seeds do not possess sufficient levels of dormancy inducing genes and chemical compounds the seeds become more susceptible to pre-harvest sprouting (PHS). This is also evident with seeds of sorghum. Earlier researcher's findings have indicated that in maize the dormancy of seeds is under the control of *vp1* gene. Fernando et al. (2001) used a PCR-based method and isolated genomic and cDNA clones from two genotypes of sorghum presenting variability with preharvest sprouting behavior and also the sensitivity towards ABA. They predicted two protein sequences with 699 amino acids which exhibited differences in the amino acid residues at positions 341 (Gly or Cys within the repression domain) and 448 (Pro or Ser). These protein sequences exhibited over 80, 70, and 60% homology to that of the proteins detected in maize, rice, and oat VP1 proteins. Further, it was observed through expression analysis that in these two lines of sorghum showed faintly a high level of *vp1* mRNA in the embryos of preharvest sprouting susceptible line rather than in the resistant line during embryogenesis. During the developmental process, it was found that these two lines exhibited variability in the timing of expression of this gene. In the line which was more susceptible to pre-harvest sprouting had the peak period of expression immediately at 20 DAP, while in the resistant line this expression peak was detected when the seeds almost completed the maturation phase. Under favorable germination conditions and in the presence of fluridone, sorghum *vp1* mRNA was found to be constantly correlated with sensitivity to ABA. This consistent correlation was not related with either the ABA content or the dormancy level.

The function of ABA and GA in shaping the dormancy level of developing sorghum (*Sorghum bicolor* [L.] Moench) seeds was studied in Redland B2 and IS 9530 sorghum varieties by Steinbach et al. (1997).

These varieties exhibited contrasting behavior in preharvest sprouting. Redland B2 was very susceptible to pre-harvest sprouting while IS 9530 was resistant. The panicles of these when sprayed with fluridone or paclobutrazol soon after fertilization so as to bring about an inhibition in the synthesis of ABA and GA has shown that, fluridone increased the germination ability of Redland B2 immature caryopses, whereas the early treatment with paclobutrazol completely inhibited germination of this variety during most of the developmental period. Further, it was observed that incubation of caryopses in 100 µM GA_{4+7} could overcome not only the inhibitory effect of paclobutrazol, but also resulted in stimulation of germination in seeds of other treatments also. The germination indices were almost zero till physiological maturity (44 DAP) in IS 9530 caryopses in paclobutrazol-treated particles as well as in untreated panicles. On the other hand, it was noticed that in fluridone-treated caryopses there was an early release of dormancy than untreated caryopses. They mentioned that their results supported the proposition that a low dormancy level (which is associated to a high preharvest sprouting susceptibility) is influenced by both low embryonic sensitivity to ABA, and high GA content.

6.2.4   GRAIN YIELD

Grain yield of *Sorghum bicolor* (L.) Moench varies with varieties and hybrids. It is well known that several morphological and physiological traits are genetically regulated and often the differences in these is observed in number of tillers, plant height, anthesis time, and many of the morphological and physiological characters. Hart et al. (2001) mapped 27 unique QTLs that regulate variation in nine morphological traits, comprising the presence versus the absence and the height of basal tillers. They described the percentage of additive genetic variance elucidated by the QTLs in a population of 137 recombinant inbred (RI) lines in two environments. Four QTLs elucidated the additive genetic variance from 86.3% to 48.9% (depending upon the environment) in the number of basal tillers with heads, and seven QTLs explained the additive genetic variance from 85.9% to 47.9% in panicle width. It is suspected that different alleles were segregating in the mapping population at any of the major dwarfing loci, but five QTLs that explained the additive genetic variance from 65.8% to 52.0% in main-culm height were mapped. They also mapped the QTLs regulating the differences in height of the tallest basal tiller, number of basal tillers per basal-tillered plant, panicle length, leaf angle, maturity, and awn length. In linkage groups (LGs) A, E, G, and I three or more QTLs were mapped, whereas none were mapped in LGs B and D. A number of the QTLs mapped in this study are probable candidates for marker-assisted selection (MAS) in breeding programs.

Quantitative traits associated with stem sugar content and grain yield characteristics are the key components governing the productivity of sorghum. In the last few decades much improvement has been attained in the improvement of sorghum crop through breeding. However, in the current years this is oriented towards the tools of genetic mapping and characterization of quantitative trait loci (QTL), for any desired trait improvement. Amukelani et al. (2010) investigated on the QTL associated with the sugar components (Brix, glucose, sucrose, and total sugar content) and sugar-related agronomic traits (flowering date, plant height, stem diameter, tiller number per plant, fresh panicle weight, and estimated juice weight) under four diverse environments using a population of 188 recombinant inbred lines (RILs) obtained by crossing grain sorghum (M71) and sweet sorghum (SS79). Using 157 AFLP, SSR, and EST-SSR markers a genetic map was made and numerous QTLs were identified by means of composite interval mapping (CIM). Further, they assessed additive × additive interaction and QTL × environmental interaction. CIM detected more than five additive QTLs in most traits elucidating the phenotypic variation range of 6.0–26.1%. A total of 24 digenic epistatic locus pairs were detected in seven traits, this confirmed the proposition that QTL analysis without considering epistasis can give rise to biased estimates. They detected QTLs which were exhibiting a numerous effects. A major QTL detected on SBI-06 was extensively allied with many traits, i.e., flowering date, plant height, Brix, sucrose, and sugar content. A major QTL × environmental interaction was observed in four out of ten studied traits.

The influence of osmotic adjustment on grain yield in *Sorghum bicolor* (L.) exposed to water stress between anthesis and maturity was investigated by Ludlow et al. (1990) in three entries (Goldrush, E57, and DK470), selected for high osmotic adjustment and other three (Texas 610SR, Texas 671, and SC 219-9-19-1) for low osmotic adjustment. They allocated these lines into early, intermediate, and late maturity groups. These entries were either well-watered, or exposed to a 50-day period of water stress after anthesis. Prior to anthesis these were well watered. At the end of the post-anthesis stress osmotic adjustment mean values of entries selected for high osmotic adjustment was more than double those selected for low osmotic adjustment. The resultant mean grain yield of entries with high osmotic adjustment was 24% higher than that of entries with low osmotic adjustment. It is observed that the entries which had more number of grains of large size resulted in higher yields. They found that the higher yield was dependent on higher harvest index and distribution index. Water stress before anthesis resulted in a larger decline in grain yields than post-anthesis stress of identical intensity. It was observed that in both the stages of water stress either the pre-anthesis or the post-anthesis, the osmotic adjustment of the entries assessed was equally effectual in lessening the decrease in grain yields.

6.3   ENVIRONMENTAL IMPACT ON SORGHUM

Boyer et al. (1982) in their analytical study of the US crops have found that in the major crops of the US, there was a huge genetic potential which was untapped. There is a requisite for realizing this genetic potential to enable the crops to adapt themselves to their environments where they were grown. Some of the evidences that were collected from the native populations of these crops have shown that there exist substantial opportunities for the improvement of productivity of these crops under unfavorable environments. Though the genotypic selection enabled in improving the agriculture, the basic mechanisms that were behind the improvement of productivity are yet poorly understood. Despite, these lacunae, the scientific advances in the recent years enable in exploring these mechanisms in much feasible manner, with larger gains in productivity.

6.3.1   WATER EXTRACTION OF SORGHUM IN SUB-HUMID ENVIRONMENT

There is an association between root growth and soil water. Based on the soil water content root growth proliferates to extract the water available in the soil. The extraction of water is analyzed by two components. Robertson et al. (1993) investigated the water extraction of grain sorghums in the sub-humid environments and the root growth relative to the rate of extraction. They analyzed the extraction in terms of components viz., (a) the time when the extraction front reached a depth. It is generally described as the moment when soil water content (θ) begins to exhibit an exponential decline with time; (b) the reduction of soil water content (θ) with time at each depth after the extraction front arrives. The second component was aimed at testing whether there would be a limitation of assimilate at the late vegetative growth due to crop shading; and whether this assimilate limitation would have its effect on the root length elongation along with the degree of root front penetration.

The results revealed that there was a constant rate of 2.72 cm day^{-1} of root front penetration downward from the time of sowing itself. It has been observed that at a depth of 100 cm down there was a lag in the extraction front than that of the root front, however, at a depth of 190 cm by which time the sorghum plant has arrived into the anthesis stage, surprisingly it was observed that both the extraction front and the root front have descended together in the soil. The sorghum plant by this time has put forth a maximum root length, and it was found that there was no shading effect either on the rate of the root front penetration or on the length of root accumulation, though there was a reduction of 70% in the above-ground growth. It was further observed that on arriving of root front to a particular soil depth, there was continuous proliferation of root, where in, the soil layer there was at least 40–20% of the extractable water.

Thus, it was noticed that root front proliferated at a faster pace rather than the extraction front. Thus, the sorghum plant has only taken a small amount of water much before the arrival of the extraction front. Consequently, the extraction rate reached maximum values at each depth of soil, only when there was retention of half the extractable water. They described the decline in soil water content (θ) by a sigmoidal curve. In the middle layers of the soil profile the description of the decline of soil water content (θ) by a sigmoidal curve was found to be better suited rather than that of an exponential curve. The sigmoidal curve is better suited in its applicability for describing the decline in soil water content with time, particularly when there is an increase in the root length with water extraction. Radio phosphorus study has indicated that in hybrid sorghum a good amount of root growth was put forth after flowering (Jerry et al., 1963).

6.3.2 *RESPONSE TO HIGHER CARBON DIOXIDE LEVELS*

In the sub-humid great plains of the USA, at Manhattan, Kansa, Chaudhuri et al. (1976) studied the effect of elevated Carbon dioxide on grain sorghum grown in a rhizotron and monitored its root and shoot growth. They have enclosed the tops of these grain sorghum plants in plastic chambers. These chambers contained 330 (ambient), 485, 660, and 795 μl l^{-1} of carbon dioxide concentration each, respectively. The results revealed a delay in the occurrence of certain stages in grain sorghum viz., boot, half bloom and soft dough stages at enriched CO_2 levels. Further, there was a higher production of shoots and roots at increased carbon dioxide levels than at ambient levels. Though number of roots and weight increased at all the soil profile depths, there was a decline in the water use per unit DM of stem, leaf, root, and grain at increased CO_2 level, particularly 795 μl l^{-1}. In addition, there was an increase in the leaf temperature and stomatal resistance, however, a lower canopy resistance was observed with an increase in the indices of leaf area.

The results of a glass house experiment enriched with elevated carbon dioxide concentration of 630 ppm with maximum temperatures of 35°C has revealed that there was a very less increase (2%) in the average carbon dioxide exchange rate (CER) in sorghum compared to other crops. Their results have shown that in determinate species such as sorghum selection of CER will not be the only criteria of selection for yield though it might hold good for indeterminate species as cotton and soybean (Mauney et al., 1977).

In subtropics there is high variability in rainfall pattern and as a result planting opportunities seem to be much limited. In these regions sorghum is taken up as a major summer crop, though the crop production appears risky. Hammer (1994) developed and tested a simulation model to assess the climatic risk on sorghum production in these environments. They identified potential limitations and tested with novel procedures. The simulation model accounted for 94% and 64% of the variations in total biomass and grain yield. The variations that were recorded with the predicted outcome of biomass and yield differences were attributed to a limitation in the prediction of harvest index.

6.4   MINERAL NUTRITION OF SORGHUM

Many of the mineral nutrient studies rely on the plants grown in nutrient solutions. The growth and development of plants in nutrient solutions is not adequate if special attention is not paid towards the preparation of the nutrient solutions. The techniques used in the growth of plants in nutrient solutions vary from crop to crop. In many of the publications the ideas are omitted and Ralph et al. (2008) in his report focused on the concerns, successes, and the experiences and problems that were faced during the growth of the sorghum plants in nutrient solutions. He discussed about the composition, pH, concentrations of P, and sources of Fe, etc., of the nutrient solutions and has given suggestions for achieving good plant growth of sorghum when grown in nutrient solutions.

A study has been undertaken by Harris (1996) on mineral application, seed priming, and sowing depth effects on emergence and growth in two varieties of sorghum viz., Segaolane, and 65D in semi-arid Region of Botswana. The results revealed that though the rate of manure application did not hasten the emergence, the growth of sorghum plants after 1st 25 DAS was enhanced. Segaolane emerged at an earlier date than 65 D and had a vigorous quick growth.

6.4.1   SEED PRIMING AND EMERGENCE OF SEEDLINGS

Seed priming is done in many crop species to improve the emergence and the plant vigor. Priming is a method where the seeds are soaked in water prior to sowing. It is one of the methods adopted to speed up the germination process. Harris (1996) has mentioned that seed priming for 0 to 10–12 h of sorghum seeds resulted in lesser period of germination time when the seeds were taken up for germination at a temperature of 30°C. An increase in the soaking period has resulted in a saving of germination time. In some seeds there was germination that continued when soaking was for 16 h and even after the stoppage of soaking, the germination continued. This has indicated that such seeds are likely to be damaged with a delay in the sowing period or when they emerge quickly on germination. As the priming time has increased for a longer time of more than 6 h, there was a hastening in emergence percentage by 23% from the soil when the temperatures were around 30C. Additionally, it was observed that the imbibition rate of the seeds was proportional to the temperature. The establishment of primed seeds was found to be successful after a rain spell or when there was slow drying of the soil after the completion of sowing. A direct proportional relationship was observed between the rate of seedling emergence and its DM production at 25 DAS. However, the vigor of individual plants and their emergence rates were dependent on the interactions that took place between the soil drying and the morphology of sorghum grain.

6.4.2 MINERAL NUTRIENT UPTAKE ENHANCEMENT BY MICROBIAL INOCULATION

Azospirillium inoculation of sorghum seeds increased the nutrient uptake of ions viz., nitrate, potassium, and phosphate by roots by two weed old root segments. The enhancement in the nutrient uptake of these ions by *Azospirillum* inoculation was more than half of the control plants uptake. This was attributed to an increase in the surface activity that is involved in these ions uptake. *Azospirillum* inoculation also resulted in some alterations in the morphology of roots, in cell arrangement of outer cortex cell layers (4–5). The effect of enhanced mineral ion uptake of inoculated plants was also evident as an increase in shoot weight of these plants by 30% in three weeks old seedlings (Lin et al., 1993).

6.4.3 GROWTH RESPONSE TO BICARBONATES AND VAM

Sorghum plants grown in nutrient solutions containing bicarbonate as $NaHCO_3$ at 0–20 mM concentration and harvested at 0, 2, 4, and 8 days of treatment were studied for its growth and uptake of other nutrients (Ramadan et al., 2008). The findings revealed that sorghum was highly sensitive to the bicarbonate concentrations and there was an early induction of chlorotic symptoms developed due to iron deficiency. An increment in the HCO_3-ion concentration in the nutrient solution, led to a reduction in the dry weights of both roots and shoots. The reduction effect of the bicarbonate ions appeared to be more pronounced with increased time of supply of these ions in the nutrient solution to the plant at higher concentrations of 10 and 20 mM. In addition, the bicarbonate ions also resulted in a drastic decline in the root length. These ions also led to marked accumulation of organic acids in the roots. The uptake rates and translocation of iron to the shoots was reduced to a larger extent and as such these bicarbonate ions in the nutrient solution induced the iron deficiency of chlorotic symptoms. The effect of bicarbonate ions also appeared in damaging the plasma membrane (PM) integrity of roots, which has resulted in an increased efflux of other ions as potassium and nitrate into the roots. The increased efflux of these ions appeared to be the main factors bringing about a decrease in root growth and impairment of iron acquisition also in the roots.

Vesicular arbuscular mycorrhizal fungi (AMF) were found to provide beneficial effects in crop plants in improving their growth rates by influencing the nutrient uptake from the soil. The Sorghum [*Sorghum bicolor* (L.) Moench] plants which were grown in growth chambers at 20, 25 and 30°C in a low P Typic Argiudoll (3.65 µg P g^{-1} soil, pH 8.3) on inoculation with vesicular-arbuscular mycorrhizal fungi (VAMF) species viz., *Glomus fasciculatum, Glomus intraradices*, and *Glomus macrocarpum* has shown that the root colonization of these species in sorghum roots were temperature dependent to a large extent. A higher degree of root colonization was observed with *G. macrocarpum*, with an increase in the plant growth and mineral uptake at a temperature of 30°C, though there was an increase in the shoot growth at two temperatures of 20 and 25°C by *G. fasciculatum* a good rate of mineral uptake was found to occur only at a temperature of 20°C. On the other hand, a depression in shoot growth along with mineral uptake was found with the inoculation of *G. intraradices* at 30°C. Among all these species, in sorghum the *G. macrocarpum* species was more effectual in improving the shoot growth, the uptake rates of phosphorus, potassium, and zinc at all the temperatures and micronutrient iron at a temperature of 25°C. These increase uptake rates of nutrients resulted in attaining an enhancement in the plant biomass. Therefore, the results have specified that the responses of sorghum plants to various species of vesicular arbuscular mycorrhizal species has to be evaluated further at varied temperature ranges to attain the beneficial effects (Raju et al., 1990).

6.5 DEFICIENCY SYMPTOMS

Clark et al. (2008) determined the effects of Mn, Fe, Zn, Cu, and B deficiencies and Mn, Fe, Zn, Cu, B, Mo, Al, Cr, Co, Ni, Se, Sr, Hg, Pb, Cd, and Ba surpluses on visual symptoms and mineral concentrations in Sorghum [*Sorghum bicolor*(L.) Moench] plants that were grown in nutrient solutions. It was found that each mineral nutrient deficiency inhibited plant growth of sorghum expect in those plants which were grown in nutrient solutions lacking B, Cu, and Fe. Distinctive visual symptoms of either deficiency or excess were manifested in sorghum by each nutrient element deficiency, however, the symptoms of excess Mo and Se were similar to those of P deficiency, and symptoms of excess Al, Cu, Co, and Ni were similar to those of Fe deficiency. Either the nutrient stresses created as deficiencies or excess affected the nutrient uptake rates and their concentrations by the plants. The nutrient concentration rates of roots were extensively affected rather than the nutrient concentrations in the leaves.

Raymond and Lockman (1972) analyzed the plant nutrient levels at four growth stages and plotted them against yield measurements. The diagrams that were developed enabled in estimating the normal nutrient levels at each growth stage of sorghum. They discussed on the climatic conditions effects on these nutrient levels and grain yields.

6.6 MANAGEMENT OF MINERALS

Kidambi et al. (1993) determined the mineral concentrations of forage sorghum for its commercial forage use of leaves and stems grown in two management systems. It was observed that higher concentration of nutrients of Ca, Mg, P, and Mn were found in the leaves than in the stems under both management systems, whereas the concentration of K, Na, and the K/ (Ca + Mg) ratio was low. Copper and Zn concentrations were not constant for either harvest schedule or leaves and stems. It was found that minerals P and Zn were most probable to be limiting for beef cattle (*Bostaurus*), but P would usually be sufficient if cattle had daily access to leafy growth. Calcium tended to be low and the K/(Ca + Mg) ratio high in stems of the hay harvests. The levels of the other minerals seemed to meet the minimum needs of beef cattle for most of the growing season. On the whole, the 10 entries of forage sorghum seemed alike in meeting mineral requirements of beef cattle.

6.7   RESEARCH ADVANCES IN INCREASING CROP PRODUCTIVITY

6.7.1   MAINTENANCE AND GROWTH RESPIRATION MODELS

Mc Cree (1974) has developed some equations as functions of photosynthetic rate, temperature, and dry weight so that they can be utilized in computer models of photosynthesis and respiration of crops. These equations were work out based on the experimental data collected on the CO_2 exchange rates of whole plants that were grown under stable conditions. They separated the dark respiratory rate into two components viz., the "maintenance" component-taken as the efflux of CO_2 after more than 48 hours in the dark at constant temperature. They found that the maintenance component was proportional to the dry weight of the plant (W). It was also a function of temperature (T). The proportionality constant (c) was less for sorghum (*Sorghum bicolor* L.), at the same temperature than clover. The second component was the "growth" component which was considered as the difference between the "maintenance" component and the total efflux during a normal night period (N). The growth component was found to be proportional to the total influx during the previous daytime period (D), and the proportionality constant (k) was not dependent on species and temperature. Their values of c and k were in conformity with values computed by Penning de Vries, utilizing the principles and equations of biochemistry.

Ronnie et al. (1997) used SORKAM, a sorghum growth model, to develop guidelines for replanting grain sorghum. They accomplished the validation of this model by using 19 sets of field data that represented 11 yrs and six locations in Kansas. By various non-parametric tests they compared the observed and simulated yields, yield components, and phenological dates and also determined the sensitivities. However, the phonological predictions by this model were sufficiently adequate; it could only capture 27 to 79% of grain yield variability at the tested locations. The yield predictions of the model from different plant populations within a planting date appeared erroneous. Further, the validation and sensitivity analysis has shown that predictions of poor yields resulted from the inappropriate computation of tiller number and faulty partitioning of biomass to caryopsis weight. Partitioning errors translocated sufficient assimilate from culm to grain produce constant yields across populations within a planting date. Thus, they specified that usage of SORKAM model for generating replant guidelines requires the improvements in modeling the associations among the various yield components and the source-sink association that as certain caryopsis weight.

Arkin et al. (1976) developed a dynamic grain sorghum model to assess the grain sorghum growth characteristics which differ very little over large regional areas in the United States, owing to their insensitivity to photoperiod and narrow genetic base among US varieties within a maturity class. The model simplified the sorghum growth attributed and enabled it to be used over large areas across the US with only minor alterations. The model is a practical approach and enables in the calculation of the daily growth and development of an average grain sorghum plant in a field stand. The growth characteristics developed in the model were emergence of leaves, their growth rate, and the timing of these events. The physical and physiological processes of light interception, photosynthesis, respiration, and water use were modeled separately and utilized as sub-models in the growth model. The accumulated dry weight for a crop is the product of the plant population and the weight of the modeled "average" plant. Most of the equations describing these processes were empirically obtained from field measurements.

Robertson et al. (1993) developed and described a simple model to describe the growth of sorghum roots. This model is based on the CERES modeling approach. The model was developed in such a way that it is possible to simulate the rooting depth and the root length density (RLD) of sorghum plant at each layer of soil. The model was analyzed for grain sorghum growing under soil drying. It had five components viz., daily growth of root length is proportionate to above-ground biomass growth; the root front descends at a constant rate from sowing till early grain-filling; daily growth of root length in well-watered conditions is partitioned among the occupied soil layers in an exponential pattern with depth; proliferation of root length is limited in any layer if the extractable soil water in that layer drops below a threshold; and a predetermined proportion of existing root length is lost due to senescence each day. The parameter values for the associations were obtained from analysis of quantified depth distributions of root length from crops of grain sorghum grown in the sub-humid subtropics of Australia, on oxisol and vertisol soil types. The soils did not show any physical or chemical limitations to root growth. Though the model simulated well the root distribution with depth, the predictions of accumulated root length were less consistent. The factor used to partition daily accumulation of root length among the occupied layers was one of the most sensitive parameters affecting the modeled distribution of root length with depth. The value of this parameter differed between well-watered and water-limited crops. The analysis revealed that it is probable to model root growth of field crops using only five simple associations, with inputs that are previously used in most crop growth models.

6.8   CONCLUSIONS

The productivity of a crop depends on various physiological and biochemical functions, mineral nutrition, biotic, abiotic factors, in addition to generic and agronomic practices used. In this chapter research advances made on various physiological functions that contribute to sorghum productivity are presented. Several factors affect different stages of crop growth starting from germination, seedling establishment, vegetative growth and flowering and fruiting stage and yield. It also deliberates various manipulation methods used to determine their influences on sorghum growth and productivity.

KEYWORDS

- **abiotic stresses**
- **concentration addition**
- **germination**
- **mineral nutrition**
- **sorghum physiology**
- **unmanned aerial vehicles**

REFERENCES

De La Rosa-Ibarra, M., & Maiti, R. K., (1994). Morphological and biochemical basis of resistance of glossy sorghum to salinity at the seedling stage. *International Sorghum and Millet Newsletter, 35,* 118–119.

De La Rosa-Ibarra, M., & Maiti, R. K., (1995). Biochemical-mechanical mechanism in glossy sorghum lines for resistance to salinity stress. *Journal of Plant Physiology, 146,* 515–519.

De La Rosa-Ibarra, M., Maiti, R. K., & Ambridge, G. L. A., (2000). Evaluation of salinity tolerance of some sorghum genotypes seedlings (*Sorghum bicolor* L. Moench). *Phyton-International Journal of Experimental Botany, 66,* 87–92.

De La Rosa-Ibarra, M., Maiti, R. K., & Quezada, M. R., (2000). Physiological and biochemical characteristics of glossy/non-glossy sorghums developed under three stress factors. *Phyton-International Journal of Experimental Botany, 68,* 1–10.

Gibson, P. C., & Maiti, R. K., (1983). Trichomes in segregating generations of *Sorghum matings*: I. Inheritance of presence and density. *Crop Science, 23,* 73–75.

Lopej-Dominguez, U. R., Maiti, R. K., & Francisco, L. C., (1996). Nutritional quality of 15 glossy sorghum forage at different growth stages in irrigated and rainfed situations. *International Sorghum and Pearl Millet Newsletter,* 72–74.

Maiti, R. K., (1983). Studies on growth and development of panicles and grains of some sorghum hybrids and their parents. In: Malik, C. P., (ed.), *Advances in Plant Reproductive Physiology* (pp. 81–112). Kalyani Publishers, New Delhi.

Maiti, R. K., & Gibson, P. T., (1980). Trichomes in segregating generations: II. Association with shoot fly resistance. *Crop Science, 23,* 76–79.

Maiti, R. K., & Moreno, L. S., (1994). A technique for evaluating sorghum lines for adaptation to dry sowing in semi-arid tropics. *International Sorghum and Millet Newsletter, 35,* 122.

Maiti, R. K., & Moreno, L. S., (1995). Seed imbibition and drying as a technique in evaluating sorghum for adaptation to dry sowing in the semi-arid tropics. *Experimental Agriculture, 31,* 57–63.

Maiti, R. K., & Trujillo, J. J. G., (1996). Some morphophysiological characters in relation to shoot fly (*Atherigona soccata* Rond.) resistance in sorghum (*Sorgum bicolor*) L. Moench. *Publicaciones Biológicas (Biological Publications), 6,* 155–158.

Maiti, R. K., (1994). Estimation of chlorophyll and epicuticular wax of some glossy and nonglossy sorghum at the seedling stage. *International Sorghum and Millet Newsletter, 35,* 119.

Maiti, R. K., (1996). *Sorghum Science* (p. 352). Science Publishers, Lebanon, USA and Oxford & IBH Co Pvt. Ltd., New Delhi, India.

Maiti, R. K., (1998). Research strategy for multiple resistances in sorghum for Northeast Mexico. *Proc. Primer International Symposium of Sorghum* (pp. 63–77). Rio Bravo INIFAF.

Maiti, R. K., Bidinger, F. R., Seshu, R. K. V., Gibson, P., & Davies, J. C., (1980). *Nature and Occurrence of Trichomes in Sorghum Lines with Resistance to Shoot Fly* (pp. 1–40). Joint progress report, Sorghum Physiology/Sorghum Entomology.

Maiti, R. K., De, L. R. M. I., & Sandoval, N. D., (1994). Genotypic variability in glossy sorghum lines for resistance to drought, salinity, and temperature stress at the seedling stage. *Journal of Plant Physiology, 143,* 241–244.

Maiti, R. K., Flores, C. L. O., & Lopej, D. U. R., (1994). Growth analysis and productivity of 15 genotypes of glossy sorghum for forge and grain production. *International Sorghum and Millet Newsletter, 35,* 133–134.

Maiti, R. K., Hernandez, P. J., & Martínez, L. S., (1992). Variability in leaf epicuticular wax and surface characteristics in glossy sorghum genotypes (*Sorghum bicolor* L. Moench) and its possible relation to shoot fly (*Atherigona soccata* Rond.) and drought resistance at the seedling stage. *Publicaciones Biológicas (Biological Publications), 6,* 159–168.

Maiti, R. K., Nuñez, G. A., Gonzalez, E. S., & Garcia, D. G., (1994). Variability in uptake of minerals of 21 glossy sorghum genotypes under seedling stress. *International Sorghum and Millet Newsletter, 35,* 116.

Maiti, R. K., Prasad, R. K. E., & Rao, V. K., (1992). Characterization and evaluation of glossy sorghum germplasm for some agronomic traits for their use in fodder and grain improvement. *Publicaciones Biológicas (Biological Publications), 61,* 169–172.

Maiti, R. K., Prasad, R. K. E., Raju, P. S., & House, L. R., (1984). The glossy traits in sorghum: Its characteristics and significance in crop improvement. *Field Crop Research, 9,* 279–289.

Maiti, R. K., Raju, P. S., & Bidinger, F. R., (1981). Evaluation of visual scoring for seedling vigor in sorghum. *Seed Science and Technology, 9,* 613–622.

Maiti, R. K., Ramaiah, K. V., Bisen, S. S., & Chidley, V. L., (1984). A comparative study of the haustorial development of *Striga asiatica* (L.) Kuntze on sorghum cultivars. *Annals of Botany, 54,* 447–457.

Maiti, R. K., Reddy, Y. V. R., & Rao, V. K., (1994). Seedling growth of glossy and non-glossy sorghums (*Sorghum bicolor* (L.) Moench) under water stress and non-stress conditions. *Phyton-International Journal of Experimental Botany, 55,* 1–8.

Maiti, R. K., Reyes-Garcia, A., Heredia, N. L., & Gamez-Gonzalez, H., (1998). Comparison of the protein profiles of for genotypes of glossy sorghum subjected to salinity at the seedling stage. *International Sorghum and Millet Newsletter,* 1101–1103.

Maiti, R. K., Rosa-Ibarra, M. D. L., Ambridge, G., & Laura, A., (1994). Evaluation of several sorghum genotypes for salinity tolerance. *International Sorghum and Millet Newsletter, 35,* 121.

Maiti, R. K., Verde, S. J., Martínez, L. S., & Rodríguez. A. J. A., (1991). Evaluation of some glossy sorghum strains for epicuticular wax, chlorophyll, and hydrocyanic acid content at the seedling stage. *Biological Publications, 5,* 27–30.

Maiti, R. K., Vidyasagar, R. K., & Swaminathan, G., (1984). Correlation studies between plant height and days to flowering to shoot fly (*Atherigona soccata* Rod.) resistance in glossy and nonglossy sorghum lines. *International Sorghum and Millet Newsletter, 35,* 109–111.

Mir, A. I., Maiti, R. K., Hernandez-Pinero, J. L., & Valades-Cerda, M. C., (2001). Variability among glossy and non-glossy sorghum genotypes in seedling and callus growth in response to 2,4-D treatments. *Indian Journal of Plant Physiology, 6,* 19–23.

Reddy, S. J., Maiti, R. K., & Seetharama, R., (1984). An interative regression approach for prediction of sorghum (*Sorghum bicolor*) phenology in the semiarid tropics. *Agricultural and Forest Meteorology, 32,* 323–338.

Steinbach, H. S., Benech-Arnold, R. L., & Sánchez, R. A., (1997). Hormonal regulation of dormancy in developing sorghum seeds. *Plant Physiology, 113,* 149–154.

CHAPTER 7

Research Advances in Abiotic Stress Resistance in Sorghum

ABSTRACT

Many abiotic stresses such as drought, high, cold temperature, salinity, flooding, etc., exert their influence on the growth and productivity of sorghum. Several concerted research studies were directed in studying their effects and to develop techniques that enabled in understanding physiological, biochemical, and molecular mechanisms of these stress factors. The overview of research advances that occurred in the following sections is presented.

7.1 ABIOTIC STRESSES

7.1.1 GROWTH AND DEVELOPMENT UNDER ABIOTIC STRESSES

Abiotic stresses, for example, salinity, drought affect sorghum growth, and development. Earlier researches have indicated that for the maintenance of cell expansion in the elongating leaf tissues, there should be a nonstop supply of nutrients to these regions. However, this region is very small. It is found to be located near to the leaf base. This region is much susceptible to nutrient disturbances, and a slight variation in the supply of nutrients is seen on the leaf expansion rates. A study was undertaken for investigating the effects of salinity on the nutrient elements availability, within the elongating region of sorghum (*Sorghum bicolor* [L.] Moench, cv. 'NK 265') at the growth zone of 6th leaf, has shown that the nutrients exhibit a spatial pattern of deposition under saline conditions. Under normal conditions, the nutrient elements potassium, calcium, magnesium, and sodium are seen to be deposited along the leaf growing zones. However, under saline stress conditions these spatial patterns varied. All through the zone of elongation, there was a decline in the deposition of the K concentration, which appeared to be inhibited to a large extent under salinity stress. The deposition rate was more inhibited as the distance from the leaf base increased, similar to the inhibition in the growth. Further, there was a reduction in the deposition of calcium also. The decline in calcium deposition resulted in further growth inhibition of leaf. Though the concentration of magnesium was found to be less, at the basal region of the growing zone of leaf, there was a higher accumulation of sodium. The accumulation of sodium in this region did not influence the growth inhibition as the growth is minimum at the basal part of the growing zone (Nirit et al., 1995).

A technique that is normally applied for modeling of net increase in crop dry-matter in non-stress environments is by the assumption that the amount of dry plant biomass produced is proportional to the intercepted photosynthetically active radiation (IPAR). It is often anticipated that the slope of this relationship or 'radiation-use efficiency' is constant for a non-stressed crop species. Kiniry et al. (1989) tested the stability of this slope both among and within grain-crop species viz., sorghum (*Sorghum bicolor* (L.) Moench), rice (*Oryza sativa* L.), wheat (*Triticum aestivum* L.), maize (*Zea mays* L.), and sunflower (*Helianthus annuus* L.). For sorghum and maize the means were 2.8 and 3.5 g/MJ IPAR, respectively.

7.2   DROUGHT

Vidya et al. (2003) reported about the effect of 28-homobrassinolide and 24-epibrassinolide brassinosteroid amelioration on seed germination and seedling growth of sorghum varieties, viz., CSH-14 and ICSV-745 (susceptible to water stress) and M-35-1 (resistant to water stress). There was an increase in the germination percentage in all the genotypes by the brassinosteroids. They mentioned that the promotion of growth by these is mainly associated with increased levels of free proline content and that of the soluble proteins. Further, the treatments also led to an increased activity of catalase, while they led to a reduction in the activities of both peroxidase and ascorbic acid oxidase in sorghum.

Sorghum can tolerate both arid and wet climates. It can be cultivated even on marginal lands. It is a foremost food and feed source. Development of high yielding varieties with improved resistance to pests and diseases along with improved practices of management enabled in increased grain yields worldwide. Many crop simulation models were developed as potential tools for management decisions and to improve yields. Many of these models use the soil plant atmosphere continuum as their research tools.

7.2.1   *ROOT PRODUCTION UNDER WATER STRESS*

Blum and Arkin (1984) studied in two isogenic lines of sorghum viz., early (58M) and late line (90M) at Texas the root growth as affected by water supply and the growth duration of the isogenic lines, in some installations of steel filled soil chambers having glass panels for root visibility. Prior to planting the soil was irrigated to field capacity and two-soil moisture regimes were studied viz., stress, and other irrigated. As a normal treatment irrigation was given when there was a 50% reduction in the soil moisture, while plants were not given any irrigation under stress treatment. The research showed that there is variation in the transpiration rates and leaf area production between the early and late maturing isogenic lines of sorghum. The early isogenic line had less transpiration. The transpiration demand of both the isogenic lines was met as long as the soil available water was more than 20% of the soil volume and once there was a decline in this percentage of available soil water, the transpiration was mainly governed by leaf area, where there was either a reduction in leaf area or senescence of leaves. However, the late isogenic line had a higher cumulative root length and root density per plant, the root density per unit of leaf area, was much smaller. Further, an increment in the root length density (RLD) per unit of the green leaf area on the plant was observed with an increase in the water stress conditions. They attributed this to the decline in the green leaf area of the plant. However, under irrigated treatment, there was a skewed pattern of root distribution in the soil profile, where more roots were concentrated in the shallow layers of the soil than at the deeper layers. Similarly, it was observed that root mortality was highly correlated with the plant age and the rooting depth and of the total cumulative root length that was found at the heading stage, 2/3rd's of this was subjected to root mortality.

Naoyuki and Yusuke (2005) studied setting up of cultivation method of sweet sorghum for monosodium glutamate (MSG) production on dry land in Indonesia, where the supply of raw materials has become limiting in recent times. Earlier, the possibility of cultivation in this area was verified during the rainy season. In the meantime, cultivation during the dry season is also significant as vast areas of unirrigated fields were left unused. The stem, which contains internodes, is the major product of sweet sorghum utilized as a raw material by fermentation industries. In this study, the internode characteristics of sorghum grown during dry and rainy seasons were compared by analyzing variations in growth and yielding ability. They cultivated a sweet sorghum cultivar-Wray in the rainy season from 1995 and in the dry season of 1996 in Madura Island of East Java, Indonesia. Sweet sorghum cultivated in dry season had shorter and lighter stems with less elongated internodes than plants cultivated in the rainy season. During the dry season the accumulation of sugar was slower in the stem, but they were supposed to be harvestable for a comparatively long period during 30–60 days after anthesis (DAA). Further studies of internode traits have shown that the variances in internode numbers (25%) and in individual internode length (75%) were the main reasons for the difference in stem length. The researchers opined that future studies must look at the cultivation period (sowing and ratoon crop), varieties, and planting density to set up a sweet-sorghum cultivation method that is appropriate for the dry season.

Bioenergy sorghum is mainly grown in water-deficit areas, so traits that increase plant water capture, water use efficiency (WUE), and resistance to water deficit are essential to increase productivity. A crop-modeling framework, APSIM, was used to predict the growth and biomass yield of energy sorghum and to locate potentially valuable traits for crop improvement. APSIM simulations of energy sorghum development and biomass accumulation replicated results from field experiments across multiple years, patterns of rainfall, and irrigation schemes. Modeling revealed that lengthy vegetative growth of energy sorghum increases water capture and biomass yield by ~30% when compared to short-season crops in a water-deficit production area. Besides, APSIM was extended to facilitate modeling of VPD-limited transpiration traits that reduce crop water use under high vapor pressure deficits (VPDs). The response of transpiration rate to rising VPD was modeled as a linear response until a VPD threshold was attained, at which the slope of the response declines, denoting a series of responses to VPD noticed in sorghum germplasm. Simulation results showed that in hot and dry regions of production the VPD-limited transpiration trait is most helpful where crops are subjected to prolonged periods without rainfall during the season or to a terminal drought. In these environments, slower but more effective transpiration improves biomass yield and averts or slow down the exhaustion of soil water and commencement of leaf senescence. In years with lower summer rainfall the VPD-limited transpiration responses noted in sorghum germplasm improved biomass accumulation by 20%. However, the ability to significantly reduce transpiration under high VPD conditions could increase biomass by 6% on average across all years. This study specified that the productivity and resistance of bioenergy sorghum grown in water-deficit environments could be further improved by developing genotypes with optimized VPD-limited transpiration traits and the exploitation of these crops in water-limited growing environments. The energy sorghum model and VPD-limited transpiration trait execution were made available to simulate performance in other target environments (Sandra et al., 2017).

Craufurd and Peacock (1993) investigated the effect of heat and drought stress on three early and four later flowering lines of sorghum. These lines were subjected to three drought stress treatments (early, late, and early plus late stress) in the field during the dry season at Hyderabad in India. However, the average diurnal temperature and evaporation rate was consistently high all through their experiment; the late and early as well as late stress conditions were severe, while the early stress was mild. The timing of the occurrence of the stress and its severity was found to exhibit a profound influence on the grain yields. A large percentage of reduction in grain yield (87%) in early flowering lines occurred when there was the stress in booting and flowering (late stress), though the grain yield was not affected by the identical stress treatment on vegetative plants. Similarly, a 50–60% reduction in grain yield was recorded with an increase in the duration of severe stress on vegetative plants (early plus late stress). Grain yield exhibited a strong and positive correlation with the number of grains m^{-2}. Variation in grain yield was associated with variation in total dry matter (DM) rather than with harvest index, which was only decreased when stress appeared at flowering. Treatment effects on thermal growth rates (g m^{-2} °Cd^{-1}) during the phase from booting to flowering established that growth during this phase was the major determining factor of yield potential (number of grains). They discussed the significance of phonology in studies related to drought resistance.

The effects of onset, amount, and distribution of rainfall planting date's on sorghum yield and water use need to be understood to choose appropriate planting date and cultivar. In this study, morphological, physi-ological, phenological, yield, and water use characteristics of different sorghum genotypes in response to different planting dates were inves-tigated under rain-fed conditions. Four genotypes (PAN8816 hybrid), Macia (open-pollinated variety, OPV), Ujiba, and IsiZulu (both landraces) were planted on 3 planting dates (early, optimal, and late) in a split-plot design, with planting dates as the main factor. It was found that the late crop establishment and low final emergence were associated with the low soil water at the optimal planting date. Low leaf number, canopy cover, chlorophyll content index and stomatal conductance, and hastened pheno-logical development make sorghum genotypes adaptable to low and fluc-tuating rainfall at the late planting date. This brings about low biomass and grain yields. Landraces showed stable grain yield across planting dates, while OPV and hybrid genotypes exhibited a significant reduction in grain yield in response to low water when planted late. The highest biomass and grain yield WUE were recorded at optimal planting date (30.5 and 9.2 kg·ha^{-1}·mm^{-1}) when compared to late (23.1 and 8.7 kg·ha^{-1}·mm^{-1}), and early planting dates (25.2 and 8.3 kg·ha^{-1}·mm^{-1}). Reduction in biomass and grain WUE of PAN8816 and Macia was observed in response to low soil water content, and irregular and unequal rainfall experienced during the late planting date. On the other hand, for Ujiba and IsiZulu biomass and grain WUE increased with decreasing rainfall. Cultivation of PAN8816 was recommended under low soil water conditions to maximize crop stand. Macia was proposed under optimal conditions. Ujiba and IsiZulu landraces were recommended for low rainfall areas with highly fluctuating rainfall. Repetition or modeling of genotype responses through environments different from Ukulinga is needed for comprehensive water use characterization of these genotypes (Hadebe et al., 2017).

7.2.2   WATER STRESS AND ABA CONTENT

An outstanding correspondence was observed between germinability, endog-enous ABA concentrations, and embryo sensitivity to ABA at various stages of sorghum development, in an investigation conducted by Benech et al. (1991) on the effect of water stress given intermittently during grain filling on the germinability of developing seeds of *sorghum bicolor*. Seeds developing in plants subjected to drought showed a high level of germinability much earlier in their maturation period than those of the control plants. In addition they were also less resistant to pre-harvest sprouting (PHS). There was also a very high level of ABA accumulation in the seeds of water-stressed plants in the early stages of development. There was a marked decline in the ABA content in seeds when they stopped growing. The ABA content was very low in these seeds when compared with the seeds developed on control plants at the end of the maturation period. Development under drought conditions decreased the sensitivity of the isolated embryo to exogenous ABA by about 10-folds. Their research findings have suggested the vital part of ABA acting as a key inhibitor of precocious germination. Further, the research findings have shown that variations in germinability affected by water stress during grain filling might be associated with variations in ABA pool size in the developing seed.

Analysis of the productivity rates in semi-arid tropics of sorghum, maize, and pearl millet under water shortage conditions at various stages of development (Muchow, 1989) revealed that sorghum could out yield millet but not maize in productivity. Similarly, though there was less grain production in maize under water-limited conditions; identical grain yield was noticed in millet and sorghum. Additionally, it was found that biomass production was found to be much stable rather than the grain yield under these water shortage conditions. However, there was a decline in the biomass this response in biomass decline was found to be correlated with a much fall in the radiation use efficiency rather than the radiation interception. This decline was much more pronounced particularly during the imposition of the water stress during the early vegetative growth stages. Further, the pre-anthesis mobilization of assimilates was found in sorghum but not to a large extent in the other two crops viz., maize, and millet. The results have indicated that when there was an occurrence of water shortages, there was much conservation of the harvest index rather than the biomass accumulation. There was a decrease in the harvest index only under the conditions of severe water deficits exhibiting their impact in declining the grain yields.

7.2.3   GROWTH UNDER DROUGHT AND QTLS

Drought is a critical agronomic problem. It is the single largest factor affecting crop yields. This setback may be lessened by developing crops that are well adapted to dry-land environments. Sorghum (*Sorghum bicolor* (L.) Moench) is one of the drought-tolerant grain crops. It is also a superb crop model for assessing the underlying mechanisms of drought tolerance. Mitchell et al. (1997) in their study developed a set of 98 recombinant inbred (RI) sorghum lines by crossing two genotypes with distinct drought reactions, TX7078 (pre-flowering-tolerant, post-flowering susceptible) and B35 (pre-flowering susceptible, post-flowering-tolerant) and characterized the RI population under water stress and well-watered conditions for the inheritance of traits related to post-flowering drought tolerance and also for the traits which were the potential constituents of grain development. Quantitative trait loci (QTL) analysis unveiled 13 genomic regions linked with one or more measures of post-flowering drought tolerance. Two QTL were spotted with major effects on yield and 'stay-green' under post-flowering drought. They observed that these loci were also related to yield under fully irrigated conditions. They suggested that these tolerance loci have pleiotropic effects on yield under non-drought conditions. Further, they also identified the Loci that were related to rate and/or duration of grain development. Their QTL analysis has shown the association of many loci with both rate and duration of grain development. It was found that the high rate and short duration of grain development were associated with larger seed size. In contrast, however, only two of these loci were found to be linked with variances in stability of performance under drought.

Plants show complex responses to drought stress. The detection of genetic factors fundamental to these complex responses enables in offering a stable base in the improvement of drought resistance. The stay-green character in sorghum (*Sorghum bicolor* L. Moench) is a post-flowering drought resistance trait, which provides resistance from premature senescence to plants under drought stress during the grain filling stage. Crasta et al. (1999) undertook a study to find QTL that are involved in regulating premature senescence and maturity traits. Additionally, they investigated the association of these QTLs under post-flowering drought stress in grain sorghum. They used a set of recombinant inbred lines (RILs) obtained from the cross B35 × Tx430 to develop a genetic linkage map. Then the map was scored for 142 restriction fragment length polymorphism (RFLP) markers. The RILs and their parental lines were assessed for post-flowering drought resistance and maturity under four environments. Simple interval mapping detected seven stay-green QTLs and two maturity QTLs. Three major stay-green QTLs (SGA, SGD, and SGG) provided 42% of the phenotypic variability (LOD 9.0) and four minor QTLs (SGB, SGI.1, SGI.2, and SGJ) significantly contributed to an additional 25% of the phenotypic variability in stay-green ratings. One maturity QTL (DFB) alone contributed to 40% of the phenotypic variability (LOD 10.0), whereas the second QTL (DFG) considerably contributed to an additional 17% of the phenotypic variability (LOD 4.9).

Composite interval mapping (CIM) verified the above findings with a further analysis of the QTL × Environment interaction. About 90% and 63% of genetic variability with heritability estimates of 0.72 for stay-green and 0.90 for maturity was explained by the identified QTLs for stay-green and maturity traits, respectively. Though stay-green ratings had a significant correlation (r=0.22, P≤0.05) with maturity, six of the seven stay-green QTLs did not show any association with the QTLs influencing maturity. Likewise, one maturity QTL (DFB) was independent of the stay-green QTLs. However, one stay-green QTL (SGG), located in the vicinity of a maturity QTL (DFG), and all markers in the vicinity of the independent maturity QTL (DFB) were significantly (P≤0.1) correlated with stay-green ratings, confounding the phenotyping of stay-green. The molecular genetic analysis of the QTLs affecting stay-green and maturity, accompanied by the association between these two inversely associated traits, provided an insight into fundamental physiological mechanisms and demonstrated that there is the probability of enhancing drought resistance in plants by pyramiding the favorable QTLs (Crasta et al., 1999).

It is seen that global sorghum production is affected by constraint drought. In sorghum, drought stress at both pre-flowering and post-flowering stages has been characterized. The drought stress at these stages results in a severe decline in grain yield. This problem is further aggravated by lodging in post-flowering drought stress. Often, it results in a complete loss of crop yields in mechanized agriculture. Kebede et al. (2001) conducted a study to find QTL regulating post-flowering drought tolerance (stay green), pre-flowering drought tolerance and lodging tolerance in sorghum using an F7 RIL population obtained by crossing SC56×Tx7000. They evaluated the RIL lines, with their parents, for the above under diverse environments. A RFLP map, which covers 1,355 cM and contains 144 loci, detected nine QTLs over seven LGs for stay green in several environments using CIM method. A comparison of the QTL locations with the previous results showed that three QTLs found on LGs A, G, and J were consistent. They deliberated it as significant since the stay green line SC56 used in their study was from a different source compared to B35 that was used in all the previous studies. Similarly, the Comparative mapping showed that two stay green QTLs found in their research resembled to stay green QTL regions in maize. Further, the reported genomic regions were found to be consistent with other agronomic and physiological traits associated with drought in maize and rice, indicating that these syntenic regions might be holding a cluster of genes with pleiotropic effects concerned with several drought tolerance mechanisms in these grass species. Furthermore, three, and four major QTLs accountable for lodging tolerance and pre-flowering drought tolerance, respectively, were identified. This study unveiled the significant and consistent stay green QTLs in various stay green sources that can logically be targeted for positional cloning. The identification of QTLs and markers involved in pre-flowering drought tolerance and lodging tolerance could assist plant breeders in manipulating and pyramiding those traits together with stay green to increase drought tolerance in sorghum.

7.2.4   DROUGHT RESISTANCE

Grain sorghum (*Sorghum bicolor* L. Moench) is a genetically diverse cereal crop cultivated in many semiarid regions of the world. Improvement of drought tolerance in sorghum is of principal importance. An association panel containing 300 sorghum genotypes from diverse races, characteristic of sorghum worldwide, was collected for genetic studies. In this study (i) the performance of the association panel was quantified under field conditions in Kansas, (ii) the association panel was characterized for phenological, physiological, and yield characteristics that might be related with drought tolerance and (iii) genotypes with higher yield potential and stability under diverse environments were identified that could be utilized in the sorghum breeding program. Findings unveiled considerable diversity for physiological and yield characteristics such as chlorophyll content, leaf temperature, grain numbers and grain weight per panicle, harvest index and yield. Significant variances were observed for plant height, grain weight, and numbers per panicle, harvest index, and grain yield among and within races. The US elite lines recorded the highest number of grains and grain weight per panicle, whereas the guinea and bicolor races had the lowest. The highest harvest index and yield were recorded in US elite lines and the caudatum genotypes. On the whole, plant height had a negative correlation with grain weight and grain number was negatively correlated with yield. Moisture stress had a negative effect on harvest index and grain numbers of all the races. Among the races, the caudatum genotypes exhibited stable grain yield across the different environments. Thus, there was a huge variability within the association panel for physiological and yield traits that could be utilized to increase drought tolerance in sorghum (Mutava et al., 2011).

Maintenance of green leaf area at maturity (GLAM), known as stay-green, is utilized as an indicator of postanthesis drought resistance in sorghum [*Sorghum bicolor* (L.) Moench] breeding programs in the USA and Australia. The crucial issue is whether retaining green leaves under postanthesis drought improves grain yield in stay-green compared with senescent hybrids.

Drought resistance is of enormous significance in crop production. The recognition of genetic factors associated with plant response to drought stress presents a strong foundation for increasing drought tolerance. Stay-green is a drought resistance trait in sorghum (*Sorghum bicolor* L. Moench) that offers plants resistance against premature senescence under severe soil moisture stress during the post-flowering stage. Wenwei et al. (2000) in their study on sorghum by the use of a RFLP map, that was built from a RIL population, detected four stay-green QTLs, situated on three LGs. The QTLs (Stg1 and Stg2) were found on LG A, with the other two, Stg3 and Stg4, on LGs D and J, respectively. Two stay-green QTLs, Stg1 and Stg2, elucidating 13–20% and 20–30% of the phenotypic variability, respectively, were reliably detected in all trials at various locations in two years. They also identified three QTLs for chlorophyll content (Chl1, Chl2, and Chl3), which explained the phenotypic variability of 25–30% under post-flowering drought stress. These QTLs of chlorophyll content corresponded with the three stay-green QTL regions (Stg1, Stg2, and Stg3). They altogether accounted for 46% of the phenotypic variation. The genes for key photosynthetic enzymes, heat shock proteins, and an abscisic acid (ABA) responsive gene were also located in Stg1 and Stg2 regions. The spatial arrangement has indicated that LG A is very vital for drought- and heat-stress tolerance and yield production in sorghum. High-resolution mapping and cloning of the consistent stay-green QTLs may assist in the development of drought-resistant hybrids and to work out the mechanism of drought-induced senescence in plants.

Andrew et al. (2009) presented the analysis of ~730-megabase in *Sorghum bicolor* (L.) Moench genome. They were able to place ~98% of genes in their chromosomal background. They used the whole-genome shotgun sequence and validated their analysis by genetic, physical, and syntenic information. Genetic recombination is mostly limited to about one-third of the sorghum genome with gene order and density analogous to those of rice. Retrotransposon buildup in recombinationally recalcitrant heterochromatin has elucidated that the genome size of sorghum is ~75% larger than that of rice. Though gene and repetitive DNA distributions have been conserved since palaeopolyploidization ~70 million years ago, most duplicated gene sets lost one member before the divergence of sorghum-rice. Intensive evolution makes one duplicated chromosomal segment seem to be only a few million years old. About 24% of genes are grass-specific and 7% are sorghum-specific. They mentioned that the latest gene and microRNA duplications may add to drought tolerance in sorghum.

7.3 SALINITY

7.3.1 *SALINITY EFFECT ON GROWTH AND DEVELOPMENT*

Läuchli and Grattan (2007) researched the effect of salinity stress on plant growth and development. An osmotic effect is created by salinity stress and this is accountable for bringing about a decline in growth in the initial phases of plant growth. The initial osmotic stress that is created by salinity is identical to the initial responses that plants exhibit under water stress. Thus, the genotypic differences are meager during the initial phases. The later effect of salinity is seen in the form of toxicity to leaves created due to the accumulation of salts in the leaves. This effect is seen slowly. Toxic effect of salts varies in salt-sensitive and salt-tolerant genotypes, as the salt-sensitive genotype have the inability to put-off the salt accumulation in leaves to higher levels of toxicity. Many research findings have shown that crop plants in general, are highly sensitive to salts during emergence and vegetative development rather than at seed germination. Therefore, at the sensitive stages, salinity shows its effect on the root and shoot and brings about an inhibition in their growth. The studies have shown that the supplemental supply of calcium can alleviate the inhibition of root and shoot growth as it is highly related to the maintenance of the selectivity of Potassium ion by the plasma membrane (PM) rather than for the Sodium ion. As discussed earlier, plants exhibit variability in sensitivity to salinity stress. Comparatively the reproductive phase of plants is considered as less sensitive rather than the vegetative phase. In few crops like rice and wheat the salinity effects are quite variable. In wheat, the salinity stress though hastens the reproductive growth, brings about a reduction in the yield potential due to inhibition of spike development, but in rice, reverse effect is observed. Rice is more sensitive to salt stress at the reproductive phase also. Salinity results in low yields. The tillering ability of the rice plant is reduced and as a very less number of tillers is produced and in a few cultivars, it is seen that more sterile spikelets are produced on these tillers.

Under field conditions, plants do not thrive only the saline stress, instead, several stresses arise which interactively inhibit the growth and development of plants. One such stress that is seen in soils along with salinity stress is the toxicity of boron. The interactions of boron and salinity are yet to be elucidated completely.

Sorghum bicolor (L.) Moench, cv. 610 was investigated for its response to salinity in the presence of mineral and phytohormone substituted nutrient solutions. This cultivar exhibited its adaptation to growth at 300 mol m^{-3} NaCl, when it was enriched with mineral nutrients in only half-strength Hoagland's solution. Though the cultivar exhibited its optimal growth at full strength Hoagland's solution, the growth rates were lower at either the high or the low concentrations. On the other hand, there was no such response in growth rate with a modified Hoagland nutrient solutions in plants which were subjected to 150 mol m^{-3} NaCl. Increased mineral nutrient concentration with the addition of phytohormones like cytokinins and gibberellic acid (GA)-induced a similar effect on the growth rate of this cultivar at high salinity levels. Growth variations of enhancement in growth with the addition of phytohormones or increased mineral nutrient concentrations are identical, however, when there is an imbalance in the phytohormones rather than that of the mineral nutrients in Hoagland nutrient solution, it was noted that at high salinity concentration these had an effect on the growth of the cultivar and there was a limitation in growth at 300 mol m^{-3} NaCl in the presence of half-strength Hoagland solution.

Thus the changes in these mineral nutrient concentrations in the nutrient medium, not only induce a nutritional effect but it appears that it acts as a signal that is involved in the hormonal balance and would allow the growth to occur at high salinity levels also. They specified that in sorghum seedlings exposed to 300 mol m^{-3} NaCl, there was a decline in the nutrient concentrations that were caused due to salinity; as a result of variations in the nutrient concentrations, the Hoagland solution was unable to sustain growth. Thus, it is evident that an adjustment of mineral nutrients in Hoagland solution may likely to induce the endogenous synthesis of cytokinins and gibberellins in the plant, which enables in bringing about the growth. However, it has been found that during the pretreatment, if there was an addition of these phytohormones, they have resulted in inhibition of the growth process and the process of adaptation of plants. Thus, their results have indicated that a plant's response to exogenous application of phytohormones for salinity adaptation is mostly dependent on the time that has been elapsed at the start of salinization (Amazallagh et al., 1992).

Ecotypes of cultivars and varieties in many crop plants continuously emerge in nature. Artificial selection enabled many breeders to achieve notable information on salt tolerance in the past few years in many of the potential crops. Evaluation of seven Saudi local cultivars of Sorghum (Bafeel, 2014) against red sea tolerance on various physiological parameters has shown that shoot growth was reduced to a large extent rather than the root growth. They found that cultivar C_3 (mix white and red seeds) was more salt tolerant. In contrast, cultivar C_4 (whitish seeds) was found to be more salt sensitive. This criterion of tolerance or sensitivity was mostly based on their ability to germinate and shoot development.

Sweet sorghum is a drought-tolerant crop with high resistance to saline-alkaline soils, and sweet sorghum may be grown an alternate summer crop for the production of biofuel in areas with limited irrigation water. A two-year field trial was undertaken in Northern Greece to evaluate the productivity (biomass, juice, total sugar, and theoretical ethanol yields) of four sweet sorghum varieties (Sugar graze, M-81E, Urja, and Topper-76-6), one-grain sorghum variety (KN-300), and one grass sorghum variety (Susu). These varieties were cultivated in intermediate (3.2 dSm^{-1}) or in high (6.9 dSm^{-1}) soil salinity with either low (120 mm) or intermediate (210 mm) irrigation water supply (supplemented with 142–261 mm of rainfall during growth). Further, they determined the effects of soil salinity and irrigation water supply on chlorophyll content index, photosystem II quantum yield, stomatal conductance and leaf K/Na ratio. In a soil salinity of 3.2 dS m^{-1} and 6.9 dS m^{-1}, the average emergence of sorghum plants was 75,083 plants ha^{-1} and 59,917 plants ha^{-1}, respectively. The grass sorghum variety Susu was the most affected cultivar with soil salinity with the average plant emergence of 53,250 plants ha^{-1}, followed by the sweet sorghum variety Topper-76-6 with the emergence of 61,250 plants ha^{-1}. In most of the cases reduction in leaf K/Na ratio was observed with the reduction in irrigation water supply, but soil salinity did not show any significant effect on it. The sorghum plant applied with 210 mm of irrigation water had 49–88% more dry biomass, juice, and total sugar yields than the yields of sorghum applied with the 120 mm of irrigation water. The dry biomass, juice, and total sugar yields of sorghum plants grown in a soil salinity of 3.2 dS m^{-1} was 42–58% higher than the yields of sorghum plants grown in 6.9 dS m^{-1} soil salinity. The maximum ethanol yield was obtained from the sorghum plants grown in a soil salinity of 3.2 dS m^{-1} with 210 mm of irrigation water (6130 L ha^{-1}, as averaged across cultivar). The study concluded that the sufficient juice, total sugar and ethanol yields were obtained from the sweet sorghum grown in fields with a soil salinity of 3.2 dSm^{-1}, though the plants were applied with 50–75% of the irrigation water usually applied to sorghum (Vasilakoglou et al., 2011).

7.3.2 NUTRIENT ACCUMULATION IN SORGHUM UNDER SALINITY

Patrick and Andre (1990) in their study on the effect of salt stress on nutrient relations and growth responses in a sorghum cultivar under field and greenhouse conditions has found that there was a substantial reduction in the DM production of the cultivar with an increase in the soil electrical conductivity from 2.1 to 5.9 dS m^{-1}. In addition there was also a reduction in both the root and shoot dry weights in plants grown under greenhouse conditions by moderate isotonic solutions of NaCl rather than those of sodium sulfate compared to those plants under field conditions. The higher rate of inhibition in growth rates was observed by a higher concentration of –0.6 MPa of sodium sulfate than sodium chloride. Similarly, it was found that at this higher concentration of sodium sulfate there was a breakdown in the sodium exclusion from the shoots of this cultivar and a concomitant rise in shoot tissue the levels of sodium and a decline in the levels of other nutrients, particularly potassium and magnesium. This cultivar also exhibited variation in the concentration of nutrients Ca, Mg, Na, and Cl in the seminal and adventitious roots. Although the adventitious roots had a higher level of potassium, the nutrients calcium, magnesium, chloride, and sodium appeared to be at higher levels in seminal roots rather than in adventitious roots.

In monocot species, the elongation of leaf tissue is a small region found near to the leaf base. This region needs a constant supply of nutrients for the maintenance of cell expansion. This region is very much susceptible to the disturbances caused in nutrient supplies. Nirit et al. (1995) investigated the salinity effect on the nutrient element availability in this elongating region of sorghum (*Sorghum bicolor* [L.] Moench, cv. 'NK 265') leaves, where plants were subjected to 1 or 100 mol m^{-3} NaCl salinity. Distinct spatial concentration patterns in the nutrient concentrations of potassium, Ca, Mg, and Na were found along the leaf-growing zone. On exposure to salinity stress there was a change in these patterns. There was a decline not only in the concentration of K but also in the deposition rate all through the zone of elongation. It was observed that with an increase in the distance from the leaf base the inhibition in the deposition rate of K also increased. The leaf growth inhibition was also attributed to a decline in the Ca nutrient. They also found that particularly, in the basal part of the growing region, the concentration of magnesium is much lesser and there was a preferential accumulation of sodium. As near the base of the growth zone salinity has the least effect on growth, high levels of Na are not the cause of growth inhibition in this salt-affected leaf tissue.

7.4 HIGH/COLD TEMPERATURE

7.4.1 HIGH-TEMPERATURE STRESS

Sorghum (*Sorghum bicolor* L. Moench) in semiarid regions is frequently subjected to short periods of high-temperature (HT) stress during reproductive development. Prasad et al. (2008) examined the effects of heat stress in DK-28 E hybrid plants. To quantify the effects of heat stress and also to identify the sensitive stages in hybrid DK-28 E, they grew these plants in growth chambers at daytime maximum/nighttime minimum optimum temperature (OT) of 32/22°C until 29 d after sowing. After that, plants were subjected to OT or HT (40/30°C) or were reciprocally transferred at 10-d intervals (10 d before flowering, 0, 10, 20, and 30 d after flowering [DAF]) from OT to HT and vice versa. They retained these transferred plants in the new temperature treatment for 10 d before returning them to the original temperature treatment. It was found that the continuous occurrence of heat stress resulted in a delay in the emergence of the panicle, with a reduction in plant height, seed set, seed numbers, seed yield, seed size, and harvest indices. However, this continuous heat stress was found not to exhibit its influence on the leaf photosynthesis. Similarly, an exposure of these hybrid plants even for short (10-d) periods of HT stress at flowering and also at 10 d before flowering resulted in a larger decline in the seed set and seed yields. The reduction in seed yield was much higher by HT stress during post-flowering stages (10, 20, and 30 DAF) than that at the early stages of seed development.

HT reduces the percentage of seed set in sorghum (*Sorghum bicolor* [L.] Moench). The comparative sensitivity of pollen and mainly pistil and the mechanistic response that induces tolerance or susceptibility to HT were not well understood and henceforth were the main objectives of this study. In a controlled environment, the male sterile (ATx399) and fertile (RTx430) lines were subjected to 30/20°C (OT), 36/26°C (HT1), and 39/29°C (HT2) from the onset of booting to seed set. Likewise, HT stress was imposed in the field, using heat tents. Reduction in pollen germination was observed under HT stress. Higher levels of reactive oxygen species (ROS) and reduced antioxidant enzyme activity and phospholipid unsaturation were recorded in pollen that to pistil under HT. The HT of 36/26 and 39/29°C severely damaged the cell organelle damage in pollen and pistil, respectively. The percentage seed set was higher in HT-stressed pistil pollinated with optimum-temperature pollen. Direct and reciprocal crosses indicated that pollen was more sensitive with higher reductions in the percentage of seed set than pistil under HT stress. The lower temperatures also had a more negative impact on pollen than pistil. Overall, pollen was more sensitive than pistil to HT stress owing to its higher susceptibility to oxidative damage than pistil (Djanaguiraman et al., 2018).

Climate variability because of instability in temperature is a global concern that threatens crop production. The necessity to know how the germplasm variation in major crops can be exploited to assist in discovering and developing breeding lines that can tolerate and adapt to temperature instabilities is more essential than ever. Here, the genetic variation associated with responses to thermal stresses were analyzed in a sorghum association panel (SAP) representative of major races and operational groups to find single nucleotide polymorphisms (SNPs) that are linked with resistance to temperature stress in a major cereal crop. The SAP revealed wide variation for seedling traits under cold and heat stress. Genome-wide analyses detected 30 SNPs that were closely linked with traits evaluated at the seedling stage under cold stress and tagged genes that act as regulators of anthocyanin expression and soluble carbohydrate metabolism. In the meantime, 12 SNPs showed a significant association with seedling traits under heat stress and these SNPs were tagged genes that serve in sugar metabolism, and ion transport pathways. Assessment of co-expression networks for genes adjacent to the significantly associated SNPs showed complex gene interactions for cold and heat stresses in sorghum. The expression of four genes in the network of Sb06g025040, a basic-helix-loop-helix (bHLH) transcription factor were focused and validated. During cold stress up-regulation of the genes in this network was observed in a moderately tolerant line as compared to the more sensitive line.

These findings have shown the potential of genotype information for the development of temperature resistant sorghum cultivars and additional characterization of genes and their networks accountable for adaptation to thermal stresses. This information on the gene networks can be applied to the other cereal crops to get better insight into the genetic basis of resistance to temperature variations during plant developmental stages (Chopra et al., 2017).

7.4.2   LOW-TEMPERATURE EFFECT

The low-temperature effect on inducing sterility in sorghum became more evident at the late archesporial cell of pollen mother cell development when analyzed in grain sorghum (CK 60 x Tx 415 by exposing the plants to five low-temperature nights (25°C day/10°C night) by Brooking (1976). This sensitivity to low temperature lasted till the leptotene stage of the meiosis. Though there was any effect on female fertility, it was found that this sensitivity of low temperature even extended from the emergence of the ligule of the flag leaf to the elongation of the flag leaf to a length of 20 cm. This was more evident when the low-temperature period extended for 6–7 days with the conditions of 25°C day/20°C night. Though there was a normal apparent development of pollen under the treatment of low temperatures, it was however, found that at the vacuolate two-celled microspore stage, prior to the maturation there was an arrest in the development of pollen. At the stage of anthesis this sterile pollen lacked starch and had very low levels of free proline.

7.5 FLOODING/EXCESS MOISTURE

The effect of waterlogging on morphological and physiological characteristics of sorghum (*Sorghum bicolor* L. Moench) cultivars was investigated. Four sorghum cultivars, cv. Wray, Keller, Bailey (sweet cultivar) and cv. SP1 (forage cultivar) were exposed to 20 days of waterlogging at five expanded leaf stage and drained pots were maintained as the control. Among cultivars, significant differences were not seen in shoot and root biomass after 20 days of waterlogging. Reduction in leaf area (69%), plant height (30%) and youngest leaf expansion rate of all cultivars and severe reduction in SP1 (35–80%) was found. In all the cultivars flooding promoted leaf senescence. Increased allocation of biomass to shoot (increase in shoot/root) in Wray, Keller, and Bailey and promoted partitioning of biomass to root in SP1 under flooding. The emergence of the new nodal root was noticed in SP1, while the capability to uphold the root surface area by increasing the longest root length and nodal root development near the soil surface was noticed in Wray. Waterlogging conditions severely reduced photosynthetic rate, stomatal conductance and transpiration rate of sweet cultivars (65–78%). However, improved photosynthetic rate, stomatal conductance and transpiration rate were recorded in forage cultivar (56%) over the control. In Wray, the capacity to maintain root surface area and allocation of more biomass to shoot during waterlogging may uphold new growth. In SP1 the ability to uphold leaf gas exchange parameters was due to the active nodal root growth. However, there was no association between photosynthetic rate and shoot growth of sorghum under anaerobic conditions (Promkhambut et al., 2010).

Waterlogging is a major constraint to crop production and an improved knowledge of plant response is needed for the improvement of crop tolerance to waterlogged soils. A class of channel-forming proteins, Aquaporins (AQPs) plays a pivotal role in water transport in plants. An investigation was carried out to study the regulation of AQP genes under waterlogging stress and to characterize the genetic variability of AQP genes in sorghum. Two tolerant and two sensitive sorghum genotypes were subjected to waterlogging stress for 18 and 96 h. Then, transcriptional profiling of AQP genes in the nodal root tips and nodal root basal regions of genotypes was performed. The results unveiled important gene-specific patterns of genotype, root tissue sample, and time point. In some tissue samples and time point combinations, PIP2-6, PIP2-7, TIP2-2, TIP4-4, and TIP5-1 expression was differentially regulated in tolerant in comparison to sensitive genotypes. A tissue-specific role of these AQP genes in alleviating waterlogging stress was suggested by this differential response. Genetic analysis of sorghum showed that AQP genes were grouped into the same four subfamilies as in maize (*Zea mays*) and rice (*Oryza sativa*). AQP gene-based phylogeny of the haplotypes was built using SNP data from 50 sorghum accessions. Phylogenetic analysis of sorghum AQP genes positioned the tolerant and sensitive genotypes utilized for the expression study in different groups. Expression analyses indicated that chosen AQPs may have a key role in tolerance of sorghum to waterlogging stress. Additional studies are required to confirm their function and to leverage phylogenetic analyses and AQP expression data to enhance waterlogging tolerance in sorghum (Suhas et al., 2017).

7.6   TOXIC EFFECTS OF MINERALS AND METHOD OF ALLEVIATION

Heavy metals like cadmium, lead, mercury are toxic to many plants by affecting the plant metabolic and biochemical processes. The reserves stored in the seed are mobilized by hydrolytic enzymes during the process of germination; these are utilized by the developing embryo which differentiates itself as a seedling. Hydrolytic enzymes hydrolyze the reserves of carbohydrates, proteins, and fats. Saritha et al. (2008) investigated the toxicity of cadmium on the hydrolytic enzymes and seed germination of sorghum. They found that sorghum seeds could tolerate low levels of cadmium of 0.5 mM; however, there was absolute end of seedling emergence at a higher cadmium concentration of 3.0 mM. Similarly, increased concentrations of cadmium resulted in decreased activity of the hydrolytic enzymes. There was also variation in the profiles of the isozymes. There was a loss in one or more of ACP isozymes, while there was an induction of a new isozyme for enzyme protease, and a total reduction in the intensity of isozymes of alpha-amylase at an increased concentration of cadmium. Apart from the variations in the isozyme profiles of sorghum, the cadmium toxicity was also evident in bringing about a change in the density of root hairs. The investigation of enzyme assays by SEM has revealed that there was an inhibition in the reserve carbohydrate viz., starch hydrolysis or its mobilization in the form of sugars from the endosperm to the developing embryo. As a result because of the inhibitory effect of cadmium toxicity on the hydrolytic enzymes, there was a decline in the germination percentage and seedling emergence. It has resulted in higher rate of disrupted seedling emergence.

Sorghum is a dry land crop and responds to its prevalent agroecosystem. The presence of heavy metals in the soils affects the seed germination and successive seedling growth of plants. It is well known that in the life cycle of a plant a key phase of crucial importance is the seed germination. The response of the plants with increased rates of seed germination is not only confined to the genetic potential of seeds but is also governed by the prevailing environmental conditions. The initial phase of seedling emergence is mostly dependent on the presence of nutritive reserves in the seeds and the activities of the hydrolyzing enzymes involved in the mobilization of these reserves to the growth of the developing embryo. Saritha et al. (2008) investigated the activities of hydrolytic enzymes under Cd toxicity during seed germination. The quantification of the cadmium uptake by seeds indicated that embryonic axes of seeds were able to tolerate levels of cadmium levels of 0.5 mM Cd. However, at increased cadmium concentration of 3.0 mM, it was observed that there was an adverse effect on the seed germination process. This was evident with the seedling growth cessation. In addition, increased cadmium concentrations also caused a reduction in the activities of all hydrolytic enzymes. However, the isozyme profile has shown that there was a loss in one or two isozymes of ACP and a new isozyme was induced for total protease, with a decrease in the intensity of alpha-amylase isozymes. These changes in isozymes were more evident at cadmium concentration of 3.0 mM. Similarly, the cadmium concentrations had an impact on bringing about a change in the density of root hairs. In addition, to the assay results the investigations done through SEM has shown that there was an inhibition in the mobilization of the starch from the endosperm. Thus, the experimental results have shown that this decrease in the hydrolysis of the reserve carbohydrates and their inhibition as well as translocation to the embryo is the cause for the decreased rate of germination and a disrupted seedling growth.

Aluminum toxicity often arises in acidic soils, leading to a decline in the uptake rate of minerals and a decrease in the plant growth. Medeiros et al. (1994) studied the effects of different micromolar concentrations of aluminum on the uptake rate of several minerals in shoots and roots of Sorghum cv SC 283, inoculated with *Glomus* species. Mycorrhizal inoculation increased the DM production of shoots and roots. In addition, it also had a considerable impact on the nutrient concentrations of P, K, Ca, Fe, Mn, and Zn of shoot; shoot contents of P, S, K, Ca, Mg, Fe, Mn, Zn, and Cu; root concentrations of P, S, K, Ca, Mn, Zn, and Cu; and root contents of Al, P, S, K, Ca, Mg, Fe, Mn, Zn, and Cu. The VAM effects on nutrient concentrations and contents and DM generally followed the sequence of UT316 > UT143 > –VAM. The VAM isolate UT143, in particular, resulted in increased uptake of zinc, while both VAM isolates increased uptake of P and Cu in shoots and roots, and many other nutrients in shoots or roots.

7.6.1   RESPONSE TO METAL TOXICITY

Plants exhibit variability's in tolerating the toxic effects of some heavy metals by some mechanisms of chelating these metals by detoxification effects. To maintain the growth and development under these conditions they also synthesize few compounds like sugars, amino acids and organic acids which enable in the osmotic regulation of the cells. Cambraia et al. (1983) assessed the effect of aluminum in two-hybrid cultivars of sorghum in their abilities to accumulate the organic acids, amino acids, and sugars. It was observed that there was an increase in the accumulation and concentration of these compounds to a large extent in the roots of cultivar having the ability to tolerate aluminum under aluminum treatment. Among the organic acids the organic acids of transaconitate and malate were accumulated to a large extent in the tolerant cultivars rather than in the susceptible cultivars. They therefore, interpreted that the accumulation of these compounds particularly under aluminum treatment in the tolerant cultivar of sorghum is indicative of one of the detoxifying chelating mechanisms.

Aluminum (Al) toxicity reduces root growth and crop yields on acid soils worldwide. However, much quantitative information is not there on protein expression profiles under Al stress in crops. In this study, the revealing of potential Al responsive proteins from root tips of Al sensitive BR007 and Al tolerant SC566 sorghum lines was reported. An approach that makes use of iTRAQ and 2D-liquid chromatography (LC) coupled with MS/MS (2D-LC-MS/MS) was used. In BR007 and SC566 total 771 and 329 unique proteins with abundance variations of >1.5 or <0.67-fold were found. Protein interaction and pathway analyses revealed that tolerant lines had more proteins involved in the antioxidant system than in the sensitive one after Al treatment. However, reverse trends were seen for proteins concerned with lignin biosynthesis. Higher levels of ROS accumulation in root tips of the sensitive line because of reduced antioxidant enzyme activity could bring about higher lignin production and hyper-accumulation of toxic Al in cell walls. The study findings showed that activities of peroxidases and the stability between production and consumption of ROS could be significant for Al tolerance and lignin biosynthesis in sorghum (Dangwei et al., 2017).

7.6.2   ALUMINUM TOLERANCE

Detection of aluminum tolerant sorghum [*Sorghum bicolor* (L.) Moench] is essential, for its successful cultivation on acid and Al-toxic soils. A quick method for screening genotypes for Al tolerance was developed by Furlani and Clark (1981) by growing over 100 plants in the same container of comparatively small volume (50 ml/plant). Mineral element conditions that gave good differential responses for Al tolerance were 148 μmol l^{-1} Al, 64 μmol l^{-1} P, 7.4 mmol l^{-1} Ca, 1.6 mmol l^{-1} Mg, 3.9 mmol l^{-1} K, 24.7 mmol l^{-1} N (8 NO_3^-:1 NH_{4+}), light (17 hours)/dark (7 hours) temperatures of 28/23 C, and solution pH values below 4.0. The genotypes exhibited differential responses to aluminum at either high or low levels of mineral elements as well as to temperatures also. These different responses of genotypes aluminum were ascribed to a negligible effect or a severe effect of aluminum. Screening for aluminum tolerance with only a single nitrate source was not found to be effective, due to an increase in pH of the solution by nitrate thereby leading to the inactivation of the effect of aluminum in the solutions where the plants grew. Thus a better genotypic separation for aluminum was seen under a combination of the nitrogen sources of both the nitrate and ammoniacal forms rather than either of the individual sources alone. The visible symptoms of toxicity of aluminum became apparent on the roots of seedlings from 10 days of their growth in the solution. The tolerance of genotypes to aluminum can be assessed effectively on the basis of root length reductions of the seminal roots of sorghum genotypes rather than by the yields of top DM or root DM.

7.7 CONCLUSIONS

Research studies assumed to work out the effects of different abiotic stresses have been discussed. Abiotic stresses caused by temperature and water are being intensified as a result of climate change. To overwhelm this problem and meet the worldwide demand for food, tolerant crops must be developed using suitable technologies. Physiological characteristics, tolerance mechanisms, and management approaches for improved crop production must be investigated in detail to efficiently apply such technologies.

KEYWORDS

- abiotic stress
- monosodium glutamate
- recombinant inbred
- sorghum
- vapor pressure deficits
- water use efficiency

REFERENCES

Amzallagh, G. N., Lerner, H. R., & Poljakoff, A., (1992). Interaction between mineral nutrients, cytokinin and gibberellic acid during growth of sorghum at high NaCl salinity. *Journal of Experimental Botany, 43*, 81–87.

Andrew, H. P., John, E. B., & Daniel, S. R., (2009). The *Sorghum bicolor* genome and the diversification of grasses. *Nature, 45*, 457–556.

Bafeel, (2014). Physiological parameters of salt tolerance during germination and seedling growth of *sorghum bicolor* cultivars of the same subtropical origin. Saudi *Journal of Biological Sciences, 21*, 300–304.

Benech, R. L., Arnold, M., Fenner, P. J., & Edwards, (1991). Changes in germinability, ABA content and ABA embryogenic sensitivity in developing seeds of *sorghum bicolor* (L.) Moench induced by water stress during grain filling. *New Phytologist, 118,* 339–347.

Blum, A., & Arkin, G. F., (1984). Sorghum root growth and water use as affected by water supply and growth duration. *Field Crops Research, 9,* 131–142.

Brooking, I. R., (1976). Male sterility in sorghum *Sorghum bicolor* (l.) Moench induced by low night temperature timing of the stage of sensitivity. *Australian Journal of Plant Physiology, 3,* 589–596.

Cambraia, J., Galvani, E. R., Estevao, M. M., & Sant, A., (1983). Effects of aluminum on organic acid, sugar and amino acid composition of the root system of sorghum (*Sorghum bicolor* L). *Journal of Plant Nutrition, 6,* 313–322.

Chopra, R., Ratan, B., Gloria, B., John, J. G., Nicholas, X., & Zhanguo, (2017). Genome-wide association analysis of seedling traits in diverse sorghum germplasm under thermal stress. *BMC Plant Biology, 45,* 67–75.

Crasta, O. R., Xu, W. W., Rosenow, D. T., Mullet, J., & Nguyen, H. T., (1999). Mapping of post-flowering drought resistance traits in grain sorghum: Association between QTLs influencing premature senescence and maturity. *Molecular and General Genetics, 202,* 579–588.

Craufurd, P. Q., & Peacock, J. M., (1993). Effect of heat and drought stress on sorghum (*Sorghum bicolor*): II. Grain yield. *Experimental Agriculture, 29,* 77–86.

Dangwei, Z., Yong, Y., Jinbiao, Z., Fei, J., Eric, C., Theodore, W. T., Leon, V. K., & Jiping, L., (2017). Quantitative iTRAQ proteomics revealed possible roles for antioxidant proteins in sorghum aluminum tolerance. *Frontiers in Plant Science, 9,* 12–19.

Djanaguiraman, M., Perumal, R., Jagadish, S. V. K., Ciampitti, I. A., Welti, R., & Prasad, P. V. V., (2018). Sensitivity of sorghum pollen and pistil to high-temperature stress. *Plant Cell and Environment, 4,* 45–56.

Furlani, P. R., & Clark, R. B., (1981). Screening sorghum for aluminum tolerance in nutrient solutions. *Agronomy Journal, 73,* 587–594.

Hadebe, S., Sandile, T., Mabhaudhi, Tafadzwanashe, M., & Albert, T., (2017). Water use of sorghum (*Sorghum bicolor* L. Moench) in response to varying planting dates evaluated under rained conditions. *South African Water Research Commission, 9,* 56–64.

Kebede, H., Subudhi, P. K., Rosenow, D. T., & Nguyen, H. T., (2001). Quantitative trait loci influencing drought tolerance in grain sorghum (*Sorghum bicolor* L. Moench). *Theoretical and Applied Genetics, 103,* 266–276.

Kiniry, J. R., Jones, C. A. J., Otoole, C., Blanchet, R., Cabelguenne, M., & Spanel, D. A., (1989). Radiation use efficiency in biomass accumulation prior to grain filling for five-grain crop species. *Field Crops Research, 20,* 51–64.

Läuchli, A., & Grattan, S. R., (2007). Plant growth and development under salinity stress. In: *Advances in Molecular Breeding Towards Drought and Salt Tolerant Crops* (Vol. 32, pp. 1–10). Springer Publications.

Medeiros, C. A. B., Clark, R. B., & Ellis, J. R., (1994). Effects of excess aluminum on mineral uptake in mycorrhizal sorghum. *Journal of Plant Nutrition, 17,* 1399–1416.

Mitchell, R. T., Edwin, G. M., Peter, B. G., & Gebisa, E., (1997). Genetic analysis of post-flowering drought tolerance and components of grain development in *Sorghum bicolor* (L.) Moench. *Molecular Breeding, 3,* 439–448.

Muchow, R. C., (1989). Comparative productivity of maize, sorghum, and pearl millet in a semi arid tropical environment: II. Effect of water deficits. *Field Crops Research, 20,* 207–219.

Mutava, R. N., Prasad, P. V. V., Tuinstra, M. R., Kofoid, K. D., & Yu, J., (2011). Characterization of sorghum genotypes for traits related to drought tolerance. *Field Crops Research, 2,* 76–82.

Naoyuki, T., & Yusuke, G., (2005). Internode characteristics of sweet sorghum (*Sorghum bicolor* (L.) Moench) during dry and rainy seasons in Indonesia. *Plant Production Science, 8,* 601–607.

Nirit, B., Wendy, K. S., & Andre, L., (1995). Growth and development of sorghum leaves under conditions of NaCl stress: Possible role of some mineral elements in growth inhibition. *Planta, 196,* 699–705.

Patrick, B. T., & Sndre, L., (1990). Growth responses and mineral nutrient relations of salt stressed sorghum. *Crop Science, 30,* 1226–1233.

Prasad, P. V., Pisipati, V. S., Mutava, R., & Tuinstra, M. R., (2008). Sensitivity of grain sorghum to high temperature stress during reproductive development. *Crop Science, 48,* 1911–1917.

Promkhambut, A., Younger, A., Polthanee, A., & Akkasaeng, C., (2010). Morphological and physiological responses of sorghum (*Sorghum bicolor* L. Moench) to waterlogging. *Asian Journal of Plant Sciences, 9,* 183–193.

Sandra, K. T., Ryan, F. M. C., & John, E. M., (2017). Bioenergy sorghum crop model predicts VPD-limited transpiration traits enhance biomass yield in water-limited environments. *Frontiers in Plant Science, 33,* 389–396.

Saritha, V. K., & Prasad, M. N. V., (2008). Cadmium stress affects seed germination and seedling growth in *Sorghum bicolor* (L). Moench by changing the activities of hydrolyzing enzymes. *Plant Growth Regulation, 54,* 143–156.

Suhas, K., Alejandra, A., Arun, P. D., Robert, P. K., Wilfred, V., Shibu, J., & Felix, B. F., (2017). Characterization and regulation of aquaporin genes of sorghum [*Sorghum bicolor* (L.) Moench] in response to waterlogging stress. *Frontiers in Plant Science, 3,* 21–30.

Vasilakoglou, Ioannis, D., Kico, K., Nikitas, G., & Thomas, (2011). Sweet sorghum productivity for biofuels under increased soil salinity and reduced irrigation. *Field Crops Research, 9,* 78–84.

Vidya, V. B., & Seeta, R. R. S., (2003). Amelioration of osmotic stress by brassinostereoids on seed germination and seedling growth of three varieties of sorghum. *Plant Growth Regulation, 41,* 25–32.

Wenwei, X., Prasanta, K. S., Oswald, R. C., Darrell, T. R., John, E. M., & Henry, T. N., (2000). Molecular mapping of QTLs conferring stay-green in grain sorghum (*Sorghum bicolor* L. Moench). *Genome, 43,* 461–469.

CHAPTER 8

Biotic Stresses Affecting Crop Productivity

ABSTRACT

The biotic stresses such as insects and diseases, affect the growth and productivity of sorghum. Several intensive research studies were undertaken in studying their effects. Recent research advances assisted in the identification of physiological, biochemical, and molecular mechanisms of these stress factors. Overviews of research advances related to these factors are presented in this chapter.

8.1 INTRODUCTION

Conventionally, sorghum is cultivated for food, feed, and fodder requirements and now it is becoming an important bioenergy crop. Many biotic and abiotic stresses constrain sorghum production. Though it is mainly self-pollinated, the discovery of cytoplasmic male sterility (CMS) and genetic male-sterility (GMS) made cross-pollination easy in sorghum. In sorghum, the case of pollination control enabled researchers to develop both pure-line varieties (as in self-pollinated crops) and hybrids and open-pollinated populations (as in cross-pollinated crops). Comprehensive understanding of diverse end uses, nature, and intensity of various productions stresses and complete theoretic and practical knowledge of genetic and crop-breeding principles is required for the genetic improvement of sorghum. Rapid advancement in biotechnology research has provided sorghum researchers with potent tools, which complement conventional crop improvement attempts to develop improved sorghum varieties (Reddy et al., 2008).

8.2   INSECT PESTS

Sorghum is infested by more than 150 insect species, among them the major pests of sorghum are shoot fly (*Atherigona soccata* Rond.), stem borers (*Chilo partellus*), armyworms (*Mythimna separata*), aphids (*Melanaphis sacchari*), mites (*Oligonychus* spp.), midge (*Contarinia sorghicola*), and head bugs (*Calocoris angustatus*). Significant advances have been made in screening and breeding for resistance to these pests. The resistant genes for these pests are present in diverse genotypes. The resistant genes for two or more insects can be integrated into single crop variety but it is quite difficult. Host-plant resistance can be exploited to control sorghum midge, green bug, mites, aphids, and head caterpillars. However, it should be supplemented with other pest control methods (Sharma, 1993).

Details of stem borer mass rearing, field infestation and assessment of sorghum plants for resistance to this pest is provided in this paper. It was found that the formation of a dead heart due to stem borer infestation causes maximum grain yield loss and therefore should be given maximum weightage while evaluating sorghum for resistance against stem borer. Easily available indigenous ingredients are effective and cheap artificial diet for rearing stem borer on a large scale. The moth emergence was 70 to 75%, with a sex ratio of 1:1. The laboratory-reared first instar larvae were released in the field uniformly using a "bazooka" applicator, which releases equal numbers of larvae in each plant, with a carrier. Nearly 12,000 germplasm lines were screened for more than three seasons and 61 germplasm lines exhibited less susceptibility to stem borer. Four lines namely IS Nos. 5470, 5604, 8320, and 18,573 exhibited stable resistance over six environments. Factors affecting the initial establishment of the larvae were found to play a major role in the development of resistance against this pest in resistant lines. Marked differences were observed between susceptible and resistant cultivars in larval success in reaching the whorl. Antibiosis and tolerance have been identified as the key resistance mechanisms (Taneja and Leuschner, 1984).

In Asia and South and Eastern Africa sorghum is mostly infested with spotted stem borer, *Chilo partellus* (Swinhoe). Host plant resistance is a key component to control this pest under subsistence farming conditions. Therefore, here the antibiosis mechanism of resistance was studied in 20 sorghum genotypes at the seedling stage by incorporating the freeze-dried leaf powder into an artificial diet. The extent of the antibiosis mechanism of resistance against *C. partellus* in sorghum can be measured by using freeze-dried sorghum leaf powder at 12.5 g per 250 ml of the standard artificial diet or by substituting 50% of the chickpea flour in the artificial diet with sorghum leaf powder. A significant variation was seen in larval survival, larval, and pupal weights, larval, and pupal periods, percentage pupation, and adult emergence of insects fed on diets impregnated with freeze-dried leaf powder of different sorghum genotypes.

Sorghum genotypes IS 1044, IS 2123, IS 1054, IS 18573, and ICSV 714 exhibited antibiosis to *C. partellus* in the form of decreased survival and development. Principal component analysis showed that there is substantial diversity in sorghum genotypes for antibiosis to *C. partellus*. Genotypes assigned in various groups can be utilized in breeding programs to expand the basis of resistance to this pest (Kishore et al., 2006).

Shoot fly *Atherigona soccata* is a major pest of sorghum. About 8000 germplasm lines of sorghum were screened for seedling trait resistant. On the abaxial surface of the resistant lines leaf trichomes were present. These trichomed lines had distinct characters, which were apparent only in the first 3 weeks. The leaves tend to be more erect and narrower, with a yellowish-green glossy appearance, which is named as 'glossy trait.' These two characters are effective tools in selecting germplasm lines with resistance against shoot fly (Maiti and Bidinger, 1979).

Shoot fly attacks sorghum at the seedling stage. Using molecular markers linked to QTL, sorghum cultivars with enhanced shoot fly resistance can be developed. In this study 168 RILs acquired by crossing 296B (susceptible) × IS18551 (resistant) and replicated phenotypic data sets taken from four test environments were used to construct a 162 microsatellite marker-based linkage map. By means of multiple QTL mapping (MQM) 29 QTL's, four each for leaf glossiness and seedling vigor, seven for oviposition, six for dead hearts, two for adaxial trichome density and six for abaxial trichome density were detected. IS18551 donated resistance alleles for most of the QTL; but, 296B also contributed alleles for resistance in six QTL. QTL of the associated component traits were co-localized, this indicated pleiotropy or tight linkage of genes. The new morphological marker *Trit* for trichome type was linked with the major QTL for component traits of resistance. Interestingly, QTL detected in this study resembled the QTL/genes for insect resistance at the syntenic maize genomic regions, signifying the conservation of insect resistance loci between these crops. Probable candidate genes for most of the QTL were present within or adjacent to the assigned confidence intervals in sorghum. Finally, the QTL found in this analysis could make available a basis for marker-assisted selection (MAS) programs meant to improve shoot fly resistance in sorghum (Satish et al., 2009).

Antixenosis, antibiosis, and recovery resistance reaction of sorghum genotypes to sorghum shoot fly, *Atherigona soccata* was evaluated under greenhouse and field conditions. To find genotypes with stable resistance against shoot fly antixenosis for oviposition was detected under multi- and dual choice conditions for IS 1054, IS 1057, IS 2146, IS 4664, IS 2312, IS 2205, SFCR 125, SFCR 151, ICSV 700, and IS 18551 genotypes. It was not noticed under no-choice conditions. For IS 2146, IS 4664, IS 2312, SFRC 125, ICSV 700, and IS 18551 genotypes antibiosis was observed in terms of prolonged larval and pupal development, and/or larval and pupal mortality. Lower percentage of tiller dead hearts were recorded in IS 1054, IS 1057, IS 2146, IS 2205, and IS 4664 genotypes when compared with the susceptible check, Swarna. The genotypes IS 2312, SFCR 125, SFCR 151, ICSV 700, and IS 18551 showed antixenosis, antibiosis, and tolerance components of resistance. These sorghum genotypes could be utilized in breeding programs to expand the genetic base and to develop sorghum cultivars resistant to shoot fly (Kumar et al., 2008).

Host plant resistance is a key component for the management of shoot fly in sorghum. In the cultivated germplasm the levels of resistance are low to moderate. The genotypes with diverse mechanisms of resistance need to be identified to pyramid the resistance genes. Here, the antix-enosis for oviposition, antibiosis, and tolerance components of resistance were studied in a varied collection of shoot fly-resistant and -susceptible genotypes. Less number of shoot fly dead hearts was recorded in the main plants and tillers of SFCR 151, ICSV 705, SFCR 125, and, IS 18551 lines at 28 days after seedling emergence. These lines produced more number of productive tillers. Further the insects fed on these lines have shown longer larval period, lower larval survival and adult emergence when compared with the susceptible check, Swarna. It was found that the physicochemical characteristics such as leaf glossiness, trichome density, and plumule and leaf sheath pigmentation were related with resistance, whereas chloro-phyll content, leaf surface wetness, seedling vigor, and waxy bloom were related with susceptibility to shoot fly and elucidated 88.5% of the total variation in dead hearts. Step-wise regression revealed that leaf glossiness and trichome density contributed to 90.4% of the total variation in dead hearts. The direct and indirect effects, correlation coefficients, multiple, and step-wise regression analysis unveiled that the sorghum plants with leaf glossiness, trichomes on the abaxial surface of the leaf, and leaf sheath pigmentation can be utilized as marker traits to identify sorghum plants resistant to shoot fly (Dhillon et al., 2005).

The sugarcane aphid, *Melanaphis sacchari* attacks sorghum and sugarcane in many regions of Africa, Asia, Australia, the Far East, and parts of Central and South America. It acts as a vector for three persistent viruses (millet red leaf, sugarcane yellow leaf, and sugarcane mosaic viruses). Its infestation causes a decrease in diastase activity and an increase in crude fiber and carbohydrates in the grain. In this paper, the status of sugarcane aphid's research, its geographical distribution, host range, nature of damage, the magnitude of crop losses, and ecobiology in sorghum is reviewed. Several germplasm accessions, A/B- and R-lines, agronomic elite lines, hybrids, and varieties were identified as sources of resistance. The components of resistance were investigated in IS 1144C, IS 12664C, and TAM 428 lines and antixenosis for colonization/establishment was found to be predominant resistant mechanism. However, in lines IS 12609C, IS 12664C, and TAM 428 antibiosis was observed for the minimum number of days to reproduction, greater mortality, shorter longevity, and production of no or fewer nymphs. The study suggested that cultural practices, natural enemies, and chemical control can collectively prevent the sugarcane aphid from attaining the economic threshold levels (Singh et al., 2004).

The sorghum seeds contain isoinhibitors of locust and cockroach gut α-amylases. These isoinhibitors were purified by saline extraction, precipitation with ammonium sulfate, affinity chromatography on Red-Sepharose and preparative RP-HPLC on Vydac. Then, through automatic degradation of the intact reduced and *S*-alkylated proteins and by manual DABITC/PITC micro sequencing of peptides taken from enzyme digests, primary structures of isoinhibitors were determined. The inhibitors were found to be the smallest plant inhibitors of α-amylase known at present. Two amino acids viz., 47 (Slα 1) or 48 (Slα 2, STα 3) were detected in inhibitors. The three isoinhibitors exhibited 38% and 87% sequence resemblance among themselves and they had 32–81% homology with the γ-purothionins isolated from wheat endosperm (Carlos and Michael, 1991).

A synthetic *cry1Ac* gene has been transferred from *Bacillus thuringiensis* (*Bt*) to sorghum plants through particle bombardment of shoot apices.

The transformed shoot apices were used to regenerate transgenic sorghum plants via direct somatic embryogenesis by means of an intermittent three-step selection approach using the herbicide Basta. Molecular characterization on the basis of polymerase chain reaction (PCR) and Southern blot analysis revealed many insertions of the *cry1Ac* gene in five plants from three distinct transformation events. In T_1 plants, inheritance and expression of the *Bt* gene were established. Enzyme-linked immunosorbent assay revealed that the leaves with 1–8 ng per gram of accumulated *Cry1Ac* protein were mechanically wounded. Transgenic sorghum plants were screened for resistance against the spotted stem borer in insect bioassays, which revealed partial resistance to the neonate larvae of the spotted stem borer. After 5 days of infestation 60% reduction in leaf damage, 40% larval mortality, and a 36% reduction in weight of the living larvae was observed over those fed on control plants. Despite the low levels of Bt δ-endotoxin expression under the control of the wound-inducible promoter, the transgenic plants exhibited partial tolerance against the first instar larvae of the spotted stem borer (Girijashankar et al., 2005).

Transgenic Bt crops contain insecticidal crystalline proteins Cry transferred from *Bacillus thuringiensis*. These insecticidal proteins controls insects pests effectively but, the insects' pest may develop resistant to Bt crops in future. To counter this resistance, measures can be taken by investigating how Bt toxins work and how insects become resistant. In this chapter, probable mechanisms of resistance, and various approaches to deal with resistance, for example, expression of numerous toxins with distinct modes of action in the same plant, modified Cry toxins active against resistant insects, and the probable usage of Cry toxins were reviewed. These strategies should make available the methods to guarantee the effective use of Bt crops for a long time (Alejandra and Mario, 2008).

8.3   DISEASES

In sorghum, anthracnose is the most destructive disease caused by *Colletotrichum sublineolum*. In plant breeding identification of markers and genes linked with resistance to this disease are important. In this study, 242 diverse accessions of sorghum were evaluated for anthracnose resistance. They mapped eight loci linked with resistance to this disease in sorghum using 14,739 SNP markers. Based on the physical distance from linked SNP markers, genes associated with disease resistance were found in all loci except locus 8. These resistant genes consist of two NB-ARC class of R genes on chromosome 10 that were partly homologous to the rice blast resistance gene Pib, two genes related to hypersensitive response (HR): autophagy-related protein 3 on chromosome 1 and 4 harpin-induced 1 (Hin1) homologs on chromosome 8, a RAV transcription factor that is also part of R gene pathway, an oxysterol-binding protein that takes part in the non-specific host resistance, and homologs of menthone: neomenthol reductase (MNR) that catalyzes a menthone reduction to produce the antimicrobial neomenthol. The genes and markers identified in this study could be used as molecular tools to improve genetic resistance against anthracnose in sorghum (Hari et al., 2013).

In warm humid regions, sorghum crop is frequently infected with anthracnose, a foliar disease caused by the fungus *Colletotrichum graminicola*. HC 136 (susceptible to anthracnose) and G 73 (anthracnose resistant) parental lines of sorghum were utilized to look into the inheritance of anthracnose resistance in sorghum. The local isolates of *C. graminicola* cultures were inoculated on F_1 and F_2 plants. A segregation ratio of 3 (susceptible): 1(resistant) was found in F_2 generation this showed that the segregation of locus for resistance to anthracnose G 73 as a recessive trait when a susceptible cultivar HC 136 was used in a cross. The bulked segregant analysis of recombinant inbred lines (RILs) resultant from the cross HC 136 × G73 revealed that the RAPD (random amplified polymorphic DNA) marker OPJ 01_{1437} is closely associated with anthracnose resistance gene in sorghum. The parental genotypes were screened with 84 random decamer primers for polymorphism. Among these, only 24 primers were polymorphic. The RAPD marker OPJ 01_{1437} was cloned and sequenced. This sequence was used to produce specific markers called sequence characterized amplified regions (SCARs). Then using the Mac Vector program a pair of SCAR markers SCJ 01-1 and SCJ 01-2 was developed. The resistant and susceptible parents together with their respective bulks and RILs were amplified with SCAR. The results of amplification established that SCAR marker SCJ 01 is at the same loci as that of RAPD marker OPJ 01_{1437} and therefore, is associated with anthracnose resistance gene. The segregation analysis of the RILs mapped the RAPD marker, OPJ 01_{1437} at a distance of 3.26 cM apart from the locus controlling anthracnose resistance on the genetic map of sorghum. Applying BLAST program, they found that the marker has shown 100% alignment with the contig{_}3966 located on the longer arm of chromosome 8. Thus, RAPD, and SCAR markers recognized in this research can be exploited in the resistance-breeding program aimed to develop sorghum cultivars resistant to anthracnose (Monika et al., 2006).

A vital foliar disease causing yield losses in sorghum is the anthracnose. Paul et al. (2003) discussed the sustainable management strategies required to be taken up regarding sorghum anthracnose in the West and Central African regions. He mentioned that many advances have taken place with reference to this particular foliar disease and they have enabled in attaining a comprehensive understanding of the pathogenic, genotypic diversities, epidemiology, and other disease management approaches in sorghum. They further highlighted that an understanding of pathogen diversity has its role in interplaying the sustainable management strategies of anthracnose disease in sorghum.

Several strategies have been developed to control Anthracnose in sorghum, among them host plant resistance has been considered as the most effective strategy. An experiment was undertaken to find molecular markers that co-segregate with $Cg1$, a dominant gene for resistance originally located in cultivar SC748-5. BTx623 is used as one of the parents in generating RFLP and AFLP maps and BAC libraries for sorghum. $F_{2:3}$ plants obtained by crossing susceptible cultivar BTx623 and SC748-5 were screened with 98 AFLP primer combinations. Four AFLP markers that cosegregate with disease resistance were detected, of which $Xtxa6227$ mapped within 1.8 cM of the anthracnose resistance locus. Previously all four AFLP markers were mapped to the end of the sorghum linkage group (LG) LG-05. Sequence scanning of BAC clones across this chromosome revealed $Xtxp549$, a polymorphic simple sequence repeat (SSR) marker, mapped within 3.6 cM of the anthracnose resistance locus. Thirteen breeding lines obtained from crosses with sorghum line SC748-5 were genotyped to study the efficiency of $Xtxa6227$ and $Xtxp549$ for MAS. The $Xtxa6227$ and $Xtxp549$ polymorphism linked with the $Cg1$ locus was observed in twelve lines. This suggested that $Xtxp549$ and $Xtxa6227$ could be utilized in MAS and pyramiding of $Cg1$ with other genes conferring resistance to Anthracnose in sorghum (Ramasamy et al., 2009).

Sorghum plant accumulates 3-Deoxyanthocyanidins phytoalexins in response to fungal infection. Phytoalexin response and expression of genes related to defense were assessed in two cultivars differing in their reaction to $Colletotrichum$ $sublineolum$, the fungus causing the anthracnose disease in sorghum. In the incompatible interaction, restricted fungal development was observed in the host during the initial stages of pathogenesis. The resistant cultivar showed defense responses through greater and faster accumulation of phytoalexins. Further, resistant cultivar showed an earlier induction of defense-related genes encoding chalcone synthase and pathogenesis-related protein PR-10. In the compatible interaction, the pathogen was found to colonize the host by multiplying primary and secondary hyphae. Along with the quantitative and timing variances, qualitative differences were observed between the cultivars in their phytoalexin response. Therefore, the resistant cultivar accumulated a complex phytoalexin mixture, including luteolinidin and 5-methoxy luteolinidin, in response to fungal inoculation and these phytoalexins were not observed in the susceptible cultivar. Previously it was reported that these compounds have higher toxicity to fungi than other phytoalexin components in sorghum. The study established that 3-deoxy anthocyanidin phytoalexins are key mechanisms of resistance to *C. sublineolum* in sorghum (Sze-Chung et al., 1999).

Plant germplasm collections have been maintained to conserve genetic variation for exploitation in crop improvement programs. Breeding for host plant resistant make available a cost-effective approach to control diseases and stabilize crop production but pathogen populations are variable and evolving; so, the detection of new resistance sources is vital. The Mozambique sorghum collection preserved by the US National Plant Germplasm System in Griffin, Georgia was inoculated with *Colletotrichum sublineolum* and examined for anthracnose resistance during the dry and wet growing seasons. Out of 22 sorghum accessions, 12 accessions showed a resistant response in both seasons. Four resistant accessions were further examined in an anthracnose disease nursery and found to be resistant, indicating the presence of host plant resistance in these accessions against different pathotypes of the disease. In the other four accessions susceptible disease response was seen during both seasons. Six accessions exhibited a differential disease response within and between experiments indicating the influence of the environment on infection response. The anthracnose resistant accessions isolated from the Mozambique germplasm collection could be an effective source of disease resistance for sorghum breeding programs (John and Louis, 2006).

The variable nature of the pathogen and an inadequate understanding of the host/pathogen interaction have been made breeding for stable host plant resistance to Anthracnose disease difficult. To develop new cultivars with durable resistance, various sources of genetic resistance must be recognized and characterized. A study was carried out to ascertain whether various sources with anthracnose resistance have different genes for resistance. Populations developed from hybridizing resistant lines were evaluated for segregation of resistance within a family. The presence of segregation (susceptible plants) within a population unveiled different resistance genes in parents. Five different sources of resistance were found in 11 germplasm lines evaluated. Segregation ratios in resistant × susceptible F_2 populations were reliable with those of simply inherited traits. The resistance was dominant in some lines and recessive in others. The resistance sources were assessed across the environment and one source (SC748-5) exhibited resistance in all environments. Hence, a source of resilient anthracnose resistance SC748-5 was found in this research which could be exploited in breeding programs to develop sorghum cultivars resistant to anthracnose (Pushpak et al., 2005).

Sorghum cultivars grown in the tropics are sensitive to photoperiod, flowering just as or after the rains terminate flowering, grains fill, and maturity. New cultivars that flower and mature earlier in the season were developed. These cultivars provide higher grain yields when soil moisture levels are usually more favorable for grain filling. However, the early flowering often exposes developing grain to wet conditions under which it deteriorates quickly. Grain molds are a main component of the sorghum grain deterioration complex, and have become a common problem for new sorghum cultivars in temperate and tropical regions. Several fungal species have been isolated from moldy grain, the most common among them are *Fusarium,* and *Curvularia* and these species differ from *F. moniliforme* which affects young developing inflorescences and many saprophytic fungi also affects mature grains. In this paper terminology, causal agents, time of infection, predisposing factors, effects on yield and quality, control measures, resistance screening procedures and progress and suggestions for additional research mainly in resistance were reviewed (Williams and Rao, 2009).

Eight-grain mold-resistant (GMR) and eight susceptible (GMS) sorghum lines were evaluated to determine the levels of four antifungal proteins (AFPs) in mature caryopses (40–45 days after anthesis (DAA)) by means of immune blot technique. In fields with grain mold incidence, the GMR group showed high levels of sormatin, chitinases, and ribosomal inactivating proteins (RIP). In a grain mold-free environment, higher RIP and lower β-1,3-glucanase levels were observed in the GMR group than the GMS group. The GMR group constantly showed higher levels of chitinase, sormatin, and RIP in the grain mold environments than in the mold-free environment. A strong correlation was observed between AFPs and grain mold resistance. When compared to GMS lines, GMR lines induced and/or retained more AFPs due to grain mold infection pressure. The coexpression of these four AFPs may be requisite for resistance to grain mold in sorghums with non-pigmented testa (Raul et al., 1999).

The complex inheritance of grain mold resistance in sorghum made its improvement difficult. In this chapter, generation means analysis was used to ascertain the inheritance of grain mold resistance. The F_1, F_2, and backcross generations obtained by crossing 'Sureño' (a dual-purpose food grain and forage variety, resistant to grain mold) and 'RTx430' (a widely adapted inbred line, highly susceptible to grain mold) were assessed in eight diverse field environments. There were significant differences in grain mold incidence between the generations across environments. Combined analysis noticed a significant generation five-environment interaction signifying that the genotypes responded in a different way to each environment. Generation means analysis of transformed grain mold revealed additive effects in all eight environments. Epistatic effects were noticed in only two of eight environments. Combined analysis showed that higher-order interactions were essential when assessed across the environment. Therefore, the study indicated that selection in specific environments is suitable for improving resistance to mold in these environments, but it may not be effectual in imparting grain mold resistance across different environmental conditions (Rodriguez-Herrera et al., 2000).

Sorghum genotypes susceptible and resistant to grain mold were screened for two seasons in plots with overhead sprinkler irrigation on rain-free days, inoculation of panicles with mold causing fungi, and bagging of panicles. Threshed grain mold ratings (TGMRs) were recorded. There was no significant difference in TGMRs of these genotypes between overhead sprinkled and unsprinkled plots in season of abundant and frequent rainfall from flowering to grain maturity. However, higher TGMRs were recorded in sprinkled plots when rainfall was low and infrequent. In the season of frequent rainfall, grain germination was lower in sprinkled than in unsprinkled plots. Under both frequent and infrequent rainfall conditions significant differences were not observed in TGMRs between inoculated and/or bagged treatments and non-inoculated and non-bagged treatment for susceptible genotypes. Inoculation and bagging increased moldiness in resistant genotypes but not to the threshold level of susceptibility. Therefore, the study findings suggested that screening of sorghum genotypes for mold resistance without inoculation and bagging of panicles is practicable if overhead sprinkler irrigation is used from flowering to harvest (Bandyopadhyay and Mughogho, 1988).

Ten sorghum genotypes differing in phenolic compound concentrations and grain mold resistance were assessed at West Lafayette, IN, for three crop seasons. They assessed variations in phenolic compounds during seed development and the effect of these variations on grain molding. Starting from 7 DAA samples were collected for 9 weeks at 7-day intervals. The concentrations of 3-deoxyanthocyanidins, flavan-4-ols, and proanthocyanidins were determined by assaying acidified methanol extracts of the sorghum seeds. Further, the level of seed infection by mold-causing fungi was observed by placing seeds on biological media. At early stages of seed development high and equal concentrations of flavan-4-ol were noted in both mold-resistant and mold-susceptible genotypes. Later susceptible genotypes showed 67% reduction in flavan-4-ol concentration between the third and the last sampling dates. Only 20% reduction was observed in the resistant genotypes in the same period. Furthermore, higher concentrations of proanthocyanidins were found in resistant genotypes (P954255, P932062, IS15346, IS7822, and P013931) than susceptible lines throughout the season. Though there were significant differences for 3-deoxyanthocyanidins among genotypes, the existence of these pigments did not make a distinction between mold-resistant and mold-susceptible genotypes. *Alternaria, Fusarium* (particularly *F. moniliforme*), *Cladosporium,* and *Epicoccum* species were the main fungi isolated from the infected seed. The study findings revealed that the maximum incidence of seed infection by these fungi occurs between 25 and 35 DAA (Admasu et al., 1996).

A study was carried out to detect the molecular markers linked to key disease-resistance loci in sugarcane. Fifty-four different sugarcane resistance gene analogue (RGA) sequences were isolated and characterized. Ten RGAs detected from a sugarcane stem expressed sequence tag (EST) library. The primers designed to conserved RGA motifs were used to isolate the remaining 44 RGAs from sugarcane stem, leaf, and root tissue. The map location of 31 RGAs was ascertained in sugarcane and compared with the location of quantitative trait loci (QTL) for brown rust resistance. After 2 years of phenotyping, 3 RGAs generated markers that were significantly linked with resistance to rust disease. To increase the understanding of the sugarcane complex genetic structure, out of 31 RGAs, 17 RGAs were mapped in sorghum. Comparative mapping of sugarcane and sorghum unveiled syntenic localization of a number of RGA clusters. The three brown rust related RGAs were mapped on the same LG in sorghum with 2 mapping to one region and the third to a region earlier shown to contain a major rust-resistance QTL in sorghum. The research establishes the significance of using RGAs for the detection of markers linked to disease resistance loci and the importance of concurrent mapping in sugarcane and sorghum (McIntyre et al., 2005).

A comparative mapping of sugarcane and sorghum was carried out in which a sugarcane cDNA clone homologous to the maize *Rp1-D* rust resistance gene was mapped in sorghum. Multiple loci were hybridized by the cDNA probe of which one was sorghum LG E where a major rust resistance QTL was mapped earlier. Rust-resistant and susceptible progeny were selected from a sorghum mapping population and partial sorghum *Rp1-D* homologs were isolated from the genomic DNA of these progeny. Sequencing of the *Rp1-D* homologs unveiled five distinct sequence classes: three from resistant progeny and two from susceptible progeny. Products from the progeny were amplified using PCR primers exclusive to each sequence class. The results proved that the five sequence classes mapped to the same locus on LG E. Cluster analysis of these sorghum sequences and existing sugarcane, maize, and sorghum Rp1-D homolog sequences revealed that the maize Rp1-D sequence and the partial sugar-cane Rp1-D homolog were clustered with one of the sorghum resistant progeny sequence classes. However, former sorghum Rp1-D homolog sequences clustered with the susceptible progeny sequence classes. Full-length sequence information was taken from one member of a resistant progeny sequence class (Rp1-SO) and compared with the maize Rp1-D sequence and a formerly known sorghum Rp1 homolog (Rph1-2). A substantial similarity was observed between the two sorghum sequences and similarity between the sorghum and maize sequences was less. The study findings put forward a conservation of function and gene sequence homology at the *Rp1* loci of maize and sorghum and make available a basis for appropriate PCR-based screening tools for presumed rust resistance alleles in sorghum (McIntyre et al., 2004).

8.4   NEMATODES

Nematode disease surveys were conducted for groundnut, pigeon pea, chickpea, sorghum, and pearl millet in Australia, Egypt, India, Jamaica, Senegal, Sudan, Thailand, and Zimbabwe. The responses received to a questionnaire on nematode problems indicated that *Meloidogyne* spp. are internationally significant nematode pests of these crops. However, *Heterodera cajani* and *Rotylenchulus reniformis* are the main pathogens of pigeon pea in India. *Meloidogyne, Pratylenchus, and Rotylenchulus species* on the legumes, and *Hoplolaimus, Pratylenchus, Quinisulcius, and Xiphinema species* on the cereals, are strongly suspected to increase the severity of fungal diseases. Cultural practices, particularly crop rotations, were found to be the most commonly employed control measures. In Brazil, Fiji, India, and the USA, research work has been undertaken to find host resistance and various sources of resistance were found against *Meloidogyne earenaria, M. incognita, M. javanica,* and *R. reniformis.* Data on damage thresholds of major pest species is available only from Brazil, Fiji, India, and the USA (Sharma and McDonald, 1982).

A field trial was performed to study the significance of sorghum, 'Pioneer 8222' and soybean rotation in the nematode management. The field in which sorghum was grown for two years and infested with *Meloidogyne arenaria* and *Heterodera glycines*, race 4, cultivars Braxton, Centennial, Gordon, Kirby, Leflore, Ransom, and Stonewall were planted. The cultivars were also planted in plots that had been in continuous soybean production. In the two cropping systems, the plants were either not treated with nematicide or was treated with a plant application of aldicarb. The application of aldicarb did not bring about any increase in yield of cultivars except for Braxton. When compared with monoculture plots, yields of all cultivars were higher in the sorghum-soybean plots. However, the cultivars 'Kirby' and 'Leflore' had a higher yield in the monoculture system. When compared with continuous soybean yield increases in response to the sorghum-soybean system varied from 31% for 'Kirby' to 231% for 'Stonewall.' The average yield increase was 85% for all cultivars. The lowest population of *H. glycines* juveniles was observed in soil samples taken 6 weeks before harvest from the plots with 'Leflore.' The populations of *M. arenaria* juveniles were highest in continuous soybean plots with 'Leflore.' Overall, juveniles were lower in plots with the sorghum-soybean rotation than in the monoculture plots. End-of-season soil populations of

H. glycines or *M. arenaria* were not affected by the aldicarb treatment (Rodriguez-Kabana et al., 1990).

In the Cauca Valley, Colombia the plant-parasitic nematodes, *Pratylenchus zeae*, *Tylenchorhynchus martini*, and *Helicotylenchus dihystera* and *Pratylenchus zeae* are the most common nematodes associated with sorghum. They were found in 100% of the samples and averaged 744 nematodes/250 cm of soil. These nematode populations exhibited significant correlations with certain soil properties including pH, phosphorus, sodium, and calcium. However, significant correlations were not observed between plant-parasitic nematode population and cation exchange capacity, sand, silt, clay, organic matter (OM), and other elements assayed (Trevathan et al., 1985).

To study the role of soil mineral nitrogen (Nmin), native *Arbuscular mycorrhizae* (AM) and nematodes in cereal/legume rotations, field experiments were performed with pearl millet (*Pennisetum glaucum*), cowpea (*Vigna unguiculata*), sorghum (*Sorghum bicolor*), and groundnut (*Arachis hypogaea*) at four sites in Niger and Burkina Faso. At all the sites legume/cereal rotations increased the grain and total dry matter (DM) yields of cereals at harvest. In comparison with the plots under continuous cereal cultivation the higher levels of soil Nmin were observed in the topsoil of cereal plots formerly sown with legumes (rotation cereals). However, the effect of this rotation on Nmin was much larger with groundnut than with cowpea. Further, higher early AM infection rates were observed in roots of rotation cereals than continuous cereals. *Helicotylenchus* sp., *Rotylenchus* sp. and *Pratylenchus* sp. were dominant plant-parasitic nematodes noticed in all experimental plots. Consistently lower nematode densities were recorded in sorghum/groundnut cropping systems compared to continuous sorghum. Continuous groundnut plots showed the lowest nematode densities which indicated that the groundnut was a poor host for the three nematode groups. In millet/cowpea cropping systems with inherently high nematode densities, crop rotations hardly showed any effect on nematode densities showing that both crops were good hosts. These findings suggested that on the nutrient-poor Sudano-Sahelian soils, the higher levels of N_{min} and AM infection early in the season increased total DM of rotation cereals compared with continuous cereals. The site-specific magnitude of these effects may be related to the efficacy of the legume species to overwhelm nematode populations and increase plant-available N through N_2-fixation (Bagayoko et al., 2000).

The decomposition of organic matter accumulates specific compounds in the soils that may be nematicidal. Soil organic amendments are mainly bio-products and wastes from industrial, agricultural, biological, and other activities. These amendments stimulate microbial activities that are antagonistic to plant-parasitic nematodes. Amendments control plant-parasitic nematodes by improving soil structure and fertility, alternating the level of plant-resistance and through the release of nemato-toxic compounds. The mode of action of organic amendments that leads to plant disease control and stimulation of microorganisms is multifaceted and depends on the nature of the amendments (Mohammad and Abdul, 2000).

8.5   INTEGRATED PEST MANAGEMENT

Plants develop several natural products, many of which have developed to give a selective advantage against microbial attack. Current progress in molecular technology, assisted by the vast power of large-scale genomics initiatives, provided an insight into the enzymatic machinery that brings about the complex pathways of plant natural product biosynthesis. In the meantime, genetic, and reverse genetic strategies provided evidences of the significance of natural products in host defense. Now it is feasible to engineer natural product pathways in plants to enhance plant disease resistance (Richard, 2001).

In temperate and tropical conditions, cases from perennial and annual crops were applied to demonstrate the research and development methods that have promoted the use and integration of host plant resistance and biological, cultural, and chemical controls. The evidences revealed that the success of IPM depends on classical experimental methods that respond to changing constraints and discoveries. Recent advancements such as genetic engineering, semiochemicals, and bioinsecticides could assist in transforming the management of various pest complexes. Tactical models are becoming more and more valuable in predicting the requirement and timing of applied controls. In many developed countries where the ultimate goal is the reduction in usage of conventional insecticides, some outstanding developments have been made in the practical application of IPM. However, an imperative need for much more applicable research and execution remains in countries where the goal is an environmentally sound mix of non-chemical and chemical methods (Way and Fvan, 2000).

The sugarcane aphid (*Melanaphis sacchari*) is a major pest of sorghum. The most effective way of controlling this pest is the exploitation of sugarcane aphid-resistant sorghum germplasm with integrated pest management strategies. The differences in plant responses to the sugarcane aphid were characterized using a resistant line (RTx2783) and a susceptible line (A/BCK60). The global sorghum gene expression, aphid feeding behavior, and inheritance of aphid resistance were examined. With the help of RNA-seq the global transcriptomic responses to sugarcane aphids were identified in resistant and susceptible plants. These responses were compared to the expression profiles of uninfested plants at 5, 10, and 15 days post-infestation. In aphid-infested susceptible plants, the genes from several functional categories including genes associated with cell wall modification, photosynthesis, and phytohormone biosynthesis exhibited altered expression. In the resistant line, only 31 genes were differentially expressed in the infested plants comparative to uninfested plants over the same time course. The results of the electrical penetration graph (EPG) unveiled that sugarcane aphid spent about twice as long in the non-probing phase, and about a quarter of the time in phloem ingestion phase on the resistant and F_1 plants. In addition, a phloem protein 2 gene expressed in both susceptible and resistant plants early (day 5) of infestation was identified through network analysis. This gene may provide defense against aphid feeding within sieve elements. On the basis of EPG and choice bioassays between susceptible, resistant, and F_1 plants, both antixenosis and antibiosis modes of resistance were found in the resistant line RTx2783. Aphid resistance from RTx2783 segregated as a single dominant locus in the F_2 generation, which could help breeders in developing sugarcane aphid-resistant hybrids using RTx2783 as the male parent (Hannah et al., 2019).

8.6 RESEARCH ADVANCE IN BIOTIC STRESS MANAGEMENT

The reactive oxygen intermediates were produced by the microbial elicitors or attempted infection with and a virulent pathogen strain. This research reported that H_2O_2 from this oxidative burst drives the cross-linking of cell wall structural proteins and acts as a local trigger of programmed death in challenged cells. Further, it functions as a diffusible signal for the induction of genes encoding cellular protectants such as glutathione

S-transferase and glutathione peroxidase in nearby cells. Therefore, H_2O_2 from the oxidative burst plays an important role in the activating localized HR during the expression of plant disease resistance (Raimund et al., 1994).

Different seed treatment practices were evaluated for their efficiency in managing seed-borne fungi related to sorghum seeds. Significant reduction in total seed-borne fungal infections was observed in seeds treated with all the five seed treatment methods viz. hot water treatment, garlic tablet, neem leaf extract, BAU-Biofungicide, and Vitavax-200. When compared with untreated control, higher germination percentage than the national standard (>80.0%) was also found in all the treatments. The Vitavax-200 was found to be best in controlling seed-borne infection followed by garlic tablets, hot water treatment and neem leaf extract. Seed-borne infection of *B. sorghicola* and *C. lunata* was reduced over 90.0% in seeds treated with neem leaf extract; while BAU-Bio fungicide provided minimum control against total seed-borne fungal infections (Masum et al., 2006).

Witch weed (Striga spp.) is a major constraint in cereals production worldwide and it gets worse with predicted climate change. In Sub-Saharan Africa and parts of Asia it is a devastating parasitic weed. The most effective way to control Striga is the implementation of integrated management strategies. The integrated management approaches mainly depend on host plant resistance. In this study, Striga resistance is introgressed into genetic backgrounds of Tabat, Wad, and Ahmed genotypes from a resistant genotype, N13 using molecular marker-assisted backcrossing. Backcross populations BC_3S_3 were generated and genotyped with SSR and diversity arrays technology (DArT) markers. Seventeen promising backcross progenies and their parents were screened in Striga infested field. The area under striga progress curve (AUSPC) revealed a significant reduction in Striga count (920-7.5) which resulted in a 97–189% increment in yield under Striga pressure. The study shows the feasible use of MAS to develop Striga resistant lines in Sudan (Ali et al., 2016).

The root parasite *Striga hermontica* (Del.) is a foremost biotic constraint to sorghum production in Africa. It was reported that *Fusarium oxysporum* Schlecht. (Foxy 2 and PSM197) is highly virulent against this weed, host-specific, and can be mass-produced. In this experiment an effort was made to develop a suitable mycoherbicidal formulation, which reduces the amount of inoculum needed for a practical field application. Chlamydospore-rich biomass of both isolates was encapsulated in a matrix composed of durum wheat-flour, kaolin, and sucrose to make "Pesta" granules. The efficiency of these granules was tested on Striga resistant and susceptible sorghum and maize cultivars under field conditions at two locations (Samaru and Bagauda) in Nigeria. 2.0 g of "Pesta" granules of each isolate were applied per planting hole in which 5 g of Striga seed-sand (1:20 w/w) were inoculated. Both granular mycoherbicides (Foxy 2 and PSM197) controlled Striga on both susceptible and resistant maize and sorghum cultivars with the same potential. On average the formulations decreased the number of emerged Striga plants per plot by 75.3, Striga dry weight by 74.4, Striga flowers by 83.6 and crop plant infested by 64.8% when compared to the controls. For maize, significant reductions were observed in all Striga parameters, whereas for sorghum the variances were small. The efficiency of both the microherbicides was enhanced by the resistant maize and sorghum cultivars. Further, different treatment × cultivar combinations were compared. The findings unveiled that the Pesta granules × resistant cultivar combination had the highest suppressive effect on Striga, particularly for maize. There was slight improvement in performance (growth, stalk dry weight, and grain yield) of host plants sorghum. However, the combination of mycoherbicides and host plant resistance significantly reduced the emergence and flowering of Striga, thus avoiding its further distribution and infestation. Therefore, the antagonists offer a good potential for Striga control in the future when combined with a long-term integrated Striga control program (Schaub et al., 2006).

8.7 CONCLUSIONS

Several biotic stresses such as insects and diseases have their effect on the plant. The research advances that have been undertaken for understanding the effects of these stress factors were discussed in this chapter. Biotic stresses caused by insects and diseases are being intensified because of favorable conditions. To overcome this problem and meet the global demand for food, tolerant crops must be developed by means of suitable technologies. Physiological aspects, mechanisms of tolerance and management strategies for better crop production must be comprehensively investigated to effectually use such technologies.

KEYWORDS

- **biotic stress**
- **genetic male-sterility**
- **marker-assisted selection**
- **menthone: neomenthol reductase**
- **multiple QTL mapping**
- **sorghum**

REFERENCES

Alejandra, B., & Mario, S., (2008). How to cope with insect resistance to Bt toxins? *Cell, 26*, 573–579.

Ali, R., Hash, C. T., Damris, O., Elhussein, A., & Mohamed, A. H., (2016). Introgression of Striga resistance into popular Sudanese sorghum varieties using marker assisted selection. *World Journal of Biotechnology, 1*, 48–55.

Bagayoko, M., Buerkert, A. G., Bationo, A., & Römheld, V., (2000). Cereal/legume rotation effects on cereal growth in Sudano-Sahelian West Africa: Soil mineral nitrogen, mycorrhizae, and nematodes. *Plant and Soil, 218*, 103–116.

Bandyopadhyay, R., & Mughogho, L., (1988). Evaluation of field screening techniques for resistance to sorghum grain mold. *Plant Disease, 72*, 500–503.

Carlos, B., & Michael, R., (1991). A new family of small (5 kDa) protein inhibitors of insect α-amylases from seeds or sorghum (*Sorghum bicolor* (L.) Moench) have sequence homologies with wheat γ-purothionins. *FEBS Letters, 279*, 279–282.

Dhillon, M. K., Sharma, H. C., Ram, S., & Naresh, J. S., (2005). Mechanisms of resistance to shoot fly, *Atherigona soccata* in sorghum. *Euphytica., 144*, 301–312.

Girijashankar, V., Sharma, H. C., & Kiran, K., (2005). Development of transgenic sorghum for insect resistance against the spotted stem borer (*Chilo partellus*). *Plant Cell Reports, 24*, 513–522.

Hannah, M. T., Sajjan, G., Erin, D. S., Tammy, G., Nathan, A. P., Gautam, S., Joe, L., & Scott, E. S., (2019). Global responses of resistant and susceptible sorghum (*Sorghum bicolor*) to sugarcane aphid (*Melanaphis sacchari*). *Frontiers in Plant Science, 9*, 23–34.

Hari, D. U., Yi-Hong, W., Rajan, S., & Shivali, S., (2013). Identification of genetic markers linked to anthracnose resistance in sorghum using association analysis. *Theoretical and Applied Genetics, 126*, 1649–1657.

John, E. E., & Louis, K. P., (2006). Variation for Anthracnose resistance within the sorghum germplasm collection from Mozambique, Africa. *Plant Pathology Journal, 5*, 28–34.

Kishore, V. K., Sharma, H. C., & Dharma, R. K., (2006). Antibiosis mechanism of resistance to spotted stem borer, *Chilo partellus* in sorghum, *Sorghum bicolor. Crop Protection, 25*, 66–72.

Kumar, I., Siva, C., Sharma, H. C., Narasu, M., & Pampapathy, G., (2008). Mechanisms and diversity of resistance to shoot fly, *Atherigona soccata* in *Sorghum bicolor*. *Indian Journal of Plant Protection, 36,* 249–256.

Maiti, R. K., & Bidinger, F. R., (1979). A simple approach to the identification of shoot-fly tolerance in sorghum. *Indian Journal of Plant Protection*, 135–140.

Masum, M. M., Islam, S. M., & Fakir, M. G., (2006). Effect of seed treatment practices in controlling of seed-borne fungi in sorghum. *Science Research and Essays, 4,* 22–27.

Mc Intyre, C. L., Casu, R. E., Drenth, J., Knight, D., Whan, V. A., Croft, B. J., Jordan, D. R., & Manners, J. M., (2005). Resistance gene analogues in sugarcane and sorghum and their association with quantitative trait loci for rust resistance. *Genome, 48,* 391–400.

Mc Intyre, C. L., Hermann, S. M., & Casu, R. E., (2004). Homologs of the maize rust resistance gene *Rp1-D* are genetically associated with a major rust resistance QTL in sorghum. *Theoretical and Applied Genetics, 109,* 875–883.

Melake-Berhan, A., Larry, G. B., Gebisa, E., & Abebe, M., (1996). Grain mold resistance and polyphenol accumulation in sorghum. *Journal of Agricultural and Food Chemistry, 44,* 2428–2434.

Mohammad, A., & Abdul, M., (2000). Roles of organic soil amendments and soil organisms in the biological control of plant-parasitic nematodes: A review. *Bioresource Technology, 74,* 35–47.

Monika, S. K., Chaudhary, H. R., Singal, C. W., & Magill, K. S., (2006). Identification and characterization of RAPD and SCAR markers linked to anthracnose resistance gene in sorghum [*Sorghum bicolor* (L.) Moench]. *Euphytica, 149,* 179–187.

Paul, S. M., Mamourou, D., Adama, N., & Fred, W. R., (2003). Sorghum anthracnose and sustainable management strategies in West and Central Africa. *Journal of Sustainable Agriculture, 25,* 43–56.

Pushpak, J., Mehta, C. C., Wiltse, W. L., Rooney, S., Delroy, C. R. A., Frederiksen, D. E., Hess, M., Chisi, D. O., & Te, B., (2005). Classification and inheritance of genetic resistance to anthracnose in sorghum. *Field Crops Research, 93,* 1–9.

Raimund, T., Richard, D., & Chris, L., (1994). H_2O_2 from the oxidative burst orchestrates the plant hypersensitive disease resistance response. *Cell, 79,* 583–593.

Ramasamy, P., Menz, M. A., & Mehta, P. J., (2009). Molecular mapping of *Cg1*, a gene for resistance to anthracnose (*Colletotrichum sublineolum*) in sorghum. *Euphytica, 165,* 597–610.

Reddy, B. V. S., Ramesh, S., Ashok, K. A., & Gowda, C. L., (2008). *Sorghum Improvement in the New Millennium*. International Crops Research Institute for the semi-arid tropics, Patancheru, Andhra Pradesh, India. ISBN: 978-92-9066-512-0.

Richard, A. D., (2001). Natural products and plant disease resistance. *Nature, 411,* 843–847.

Rodríguez-Herrera, R., Ralph, D. W., & William, L. R., (1999). Antifungal proteins and grain mold resistance in sorghum with nonpigmented testa. *Journal of Agriculture and Food Chemistry, 47,* 4802–4806.

Rodriguez-Herrera, Rooney, W. L., Rosenow, D. T., & Frederiksen, R. A., (2000). Inheritance of grain mold resistance in grain sorghum without a pigmented testa. *Crop Science, 40,* 54–62.

Rodriguez-Kabana, R., Weaver, R., & Robertson, D. B., (1990). Sorghum in rotation with soybean for the management of cyst and root knot nematodes. *Nematropica, 20,* 111–1190.

Satish, K., Srinivas, G. R., Madhusudhana, P. G., Padmaja, Nagaraja, R., Murali, M. R. S., & Seetharama, N., (2009). Identification of quantitative trait loci for resistance to shoot fly in sorghum [*Sorghum bicolor* (L.) Moench]. *Theoretical and Applied Genetics, 119*, 1425–1433.

Schaub, B., Marley, P., Elzein, A., & Kroschel, J., (2006). Field evaluation of an integrated *Striga hermontica* management in Sub-Saharan Africa: Synergy between striga-mycoherbicides (biocontrol) and sorghum and maize resistant varieties. *Journal of Plant Diseases and Protection, 20*, 691–699.

Sharma, H. C., (1993). Host-plant resistance to insects in sorghum and its role in integrated pest management. *Crop Protection, 12*, 11–34.

Sharma, S. B., & Mc Donald, D., (1982). *Global Status of Nematode Problems of Groundnut, Pigeon Pea, Chickpea, Sorghum and Pearl Millet, and Suggestions for Future Work, 90*, 216–219. oar.icrisat.org.

Singh, B. U., Padmaja, P. G., & Seetharama, N., (2004). Management of the sugarcane aphid, *Melanaphis sacchari* (Zehntner) (Homoptera: Aphididae), in sorghum: A review. *Crop Protection, 9*, 739–755.

Sze-Chung, Clivelo, K., & Verdier, R., (1999). Accumulation of 3-deoxyanthocyanidin phytoalexins and resistance to *Colletotrichum sublineolum* in sorghum. *Physiological and Molecular Plant Pathology, 55*, 263–273.

Taneja, S. L., & Leuschner, K., (1984). Methods of rearing, infestation, and evaluation for *Chilo partellus* resistance in sorghum. *Proceedings of the International Sorghum*. www.core.ac.uk.

Trevathan, L. E., Cuarezma-Teran, J. A., & Gourley, L. M., (1985). Relationship of plant-nematodes and edaphic factors in Colombian grain sorghum production. *Nematropica, 4*, 34–42.

Way, M. J., & Fvan, E. H., (2000). Integrated pest management in practice-pathways towards successful application. *Crop Protection, 19*, 81–103.

Williams, R. J., & Rao, K. N., (2009). A review of sorghum grain molds. *International Journal of Pest Management, 8*, 200–211.

CHAPTER 9

Methods of Cultivation in Sorghum

ABSTRACT

In this chapter, the different methods of sorghum crop cultivation are presented. Sorghum cultivation includes various operations from land preparation to marketing, which are briefly discussed here. Besides this, the research advances made in sorghum cultivation are discussed.

9.1   INTRODUCTION

Sweet sorghum is a C_4 crop with juicy stalks rich in sugars, and ethanol can be produced from its stalks at much lower costs than other starchy crops like maize. Sorghum can be utilized for both thermo-electrical energy and biofuel. Increasing attention in bioenergy, especially in bioethanol, presents a big challenge for this comparatively new crop. However, the quantitative and qualitative production of sweet sorghum mainly relies on the application of suitable and improved agronomic management practices, which are still to be known in some aspects. This review attempted to collect some information on the most suitable agricultural practices for sweet sorghum, while pinpointing the weak points that need attention in future studies, particularly under temperate climates. Sweet sorghum is a row crop like maize, so the agronomic management and other well-known cultural techniques used for maize can be adapted in sorghum. When compared to other competing energy crops, low requirements of input, low cost of production, drought resistance, usefulness, and high yields makes sweet sorghum a better energy balance crop, particularly when bagasse is processed to energy. In non-traditional potential growing areas such as in temperate climates of Europe where productivity or adaptation improvements via genetically modified crops is not permitted, suitable, and sustainable agricultural techniques set up the most immediate alternative to improve yields. The scientific reports revealed that research efforts are especially required for harvesting techniques, handling, and storing (Walter et al., 2012).

9.2   CROPPING SYSTEMS

To increase precipitation-use efficiency, reduce groundwater depletion by irrigation, decrease pumping costs and conserve energy, improved cropping and residue management methods are desirable. In this study, irrigated winter wheat (*Triticum aestivum* L.) and dryland grain sorghum (*Sorghum bicolor* L.) cropping system was evaluated for water storage between wheat harvest and sorghum planting time. Grain sorghum growth and yields were also recorded. During 11 months period no-tillage (NT), sweep, and disk methods were applied between crops for the management of wheat residue and weed control. During the 11 months of fallow average precipitation stored as soil water was 35, 23, and 15%. At sorghum planting average available soil water contents to the 1.8-m depth was 21.7, 17.0, and 15.2 cm. The average sorghum grain yields were 3,140, 2,500, and 1,930 kg ha^{-1} with water-use efficiencies (WUE) of 89, 77, and 66 kg ha^{-1}cm^{-1} for the relevant treatments. Grain yields were somewhat lower than sorghum forage yields; therefore forage sorghum had more WUE than grain sorghum. However, the reduction of forage WUE decreased showed the same trend as for grain. Net returns (without land, taxes, and interest costs) for March 1978 sorghum production were two times greater with NT and four times greater with sweep tillage than with disk tillage, which is most extensively used tillage method after irrigated wheat in the Southern Great Plains (Paul and Allen, 1979).

In tropics Sorghum (*Sorghum bicolor* (L.) and soybean (*Glycine max*) are often intercropped which improves soil fertility, and helps in effective utilization of water by crop and better weed control. An experiment was undertaken to find out the effects of sorghum-soybean intercropping on soybean N_2-fixation and plant composition of both crops. Tall and semi-dwarf sorghum cultivars were intercropped with soybeans at 16 population densities from 1.38 to 33 plants/m^2. Acetylene reduction was used to estimate soybean N_2-fixation. The plant parts of the both crops were analyzed for N. in soybeans grown with tall sorghum nitrogen-fixation was decreased by 99% as a result of decrease in nodules number per plant

(77%), weight per nodule (50%), and specific nodule activity, SNA (96%). Further, 87% reduction in soybean dry matter (DM) and 8% reduction in seed oil percentage were observed. Nitrogen fixation by soybeans grown with semi-dwarf sorghum was 2.64 times more than plants in monoculture; however, it yielded 40% less DM with 3% less oil in seed. Intercropping did not show any effect on soybean percent seed protein and leaf N in both cases. The increment in N_2-fixation was possibly due to an increase in nodules number per plant (62%) and SNA (36%). Probable contribution of allelopathy, NO_3 absorption, late senescence, and inter-specific mechanical support was taken into account. Protein yield of intercropped tall and semi-dwarf sorghum was decreased by 15 and 71%, respectively. In sorghum dwarf cultivar and soybean intercropping percent, seed protein was increased by 15%. Intercropping did not show any effect on sorghum grain oil (Wahua and Mille, 1978).

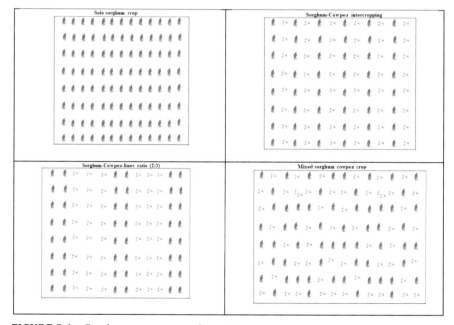

FIGURE 9.1   Sorghum cowpea cropping systems.

Twelve early maturing genotypes of grain sorghum were grown in spring and fall season and studied for the genotypic and phenotypic variances, and covariance between five characteristics. For almost all characteristics significant variances were noticed among genotypes and genotype × season interaction. Maximum phenotypic variance, attributed by genotypic variance, was found for plant height, 100-grain weight, and grain yield. Days to heading exhibited highest genotype × environment interaction. Harvest index was negatively associated with plant height and positively associated with grain yield at both phenotypic and genotypic level. This correlation analysis results could assist to select sorghum genotypes with high yield and short plant height for a double-cropping system in Japan (Nguyen and Tomohiko, 1999).

In lower midlands of Kenya variations in temperature and rainfall patterns made the area hotter and drier, which resulted in lower sorghum yields in these marginal areas. A field experiment was undertaken using cowpea-sorghum intercropping to make up for lower sorghum yields. Four intercropping patterns were arranged as treatments in randomized block design (RCBD) by replicating four times. The four treatments were sole sorghum (control), sorghum-cowpea intercropping, sorghum-cowpea-lines ratio (2:3), and mixed sorghum cowpea sowing. Starting from 7 days after emergence, the parameters like plant height, number of leaves and leaf broadness were recorded at every 13 days. The total grain yield was determined at harvest. The broader leaves and higher grain yield of sorghum (2.9 t ha^{-1}) were recorded in sorghum-cowpea intercropping. Comparatively lower yields were noted in sorghum-cowpea-lines ratio (2:3) (2.5 t ha^{-1}) and mixed sorghum cowpea sowing (2.3 t ha^{-1}). The sole crop-sorghum produced lowest yield (1.8 t ha^{-1}), which was lower than the cultivar potential of 2t ha^{-1}. The higher sorghum yield in sorghum-cowpea uniform intercropping was due to the fertility advantage of legume crop. The study findings indicated the significance of utilizing suitable intercropping patterns for realization of intercropping advantages in areas with low soil fertility (Andrew et al., 2016).

An experiment was conducted to assess the growth and yield potential of sorghum/cowpea intercrops over sole cropping pattern. Three sorghum varieties, i.e., two semi dwarfs (Pioneer and S-1007) and one tall local variety (Shahlaa) were included as treatments. Semi dwarfs varieties were planted as solid culture at the 200, 300 and 400 plants/hectare densities. The tall variety was planted at densities of 160, 240 and 320 plants/hectare. The densities for the three varieties were equal to 100, 150, and 200% of the solid recommended culture. Two rows of sorghum were planted alternating with two rows of cowpea. In sorghum-cowpea intercropping system sorghum plants showed greater potentiality and recorded higher plant height, number of leaves per plant, stem diameter, leaf area, DM accumulation, and grain yield per plant. However, intercropping pattern recorded lower grain yield per hectare than solid pattern. Therefore, the increase in grain yield per plant and plant density could not make up for the reduction of the area covered by the crop under solid culture (Refay et al., 2013).

Ghosh et al. (2004) conducted a study on deep vertisols of Bhopal, India in three cropping systems in combination of organic and inorganic fertilizers on root and shoot biomass, chlorophyll content, nodulation, and enzyme activity. The results indicated that there were a less number of nodules and nodule mass in intercrop of soybean rather than a sole soybean crop. This decline in nodule number and mass was more with higher dose of NPK. The 22.0% and 7.6% higher nodule mass were recorded in FYM treated plots than poultry manure and phosphocompost plots, respectively. In addition, organically treated plots had higher total chlorophyll content than 100% NPK plots especially at 30 days after sowing (DAS, pre-flowering). In sorghum, a peak activity of nitrate reductase was found at 60 DAS. In addition, there was a higher activity of Nitrate reductase in intercrop sorghum than that in sole sorghum and a greatest activity of this enzyme was found in 100% NPK. Further, in intercropping of soybean and sorghum there was higher production of root and shoot biomass. Similarly, there were rapid crop growth rates during 30–60 DAS. The sequence order of crop growth rates was intercropping > sole sorghum > sole soybean. A considerable increase in the DM production was found with enhancement in the dose of NPK from 0% to 100% in sole sorghum and also in the intercropping systems of soybean/sorghum.

Cereal-legumes intercropping are one of the most cost-effective and successful agronomic approaches to improve forage biomass production, nutritional quality, and economic returns. Here, the research findings on effects of intercropping on productivity, quality, competitiveness, and economic viability of sorghum-legumes mixed, row, and strip intercropping systems under varied pedo-climatic conditions were summarized. Although component crops exhibit yield reductions in row (additive and row-replacement series), mixed (seed blended crops) and strip intercropping systems, generally total productivity per unit land area rises to a large extent. This greater productivity of intercrops in temporal and spatial dimensions can be elucidated by their significantly higher resource acquiring and better utilization efficiency. Furthermore, forage intercrops bring about improved nutritional quality as protein quantity of legumes is double than cereals. Cereal-legumes intercropping systems produces higher quantities of lush green forage with improved quality traits, which eventually increase economic returns. Additionally, inclusion of legumes as an intercrop with cereals acts as a nitrogen-saving strategy owing to their biological nitrogen fixation (BNF) process. Also, cereal-legume intercropping systems provides extended soil cover and helps in reducing weed infestations and soil erosion, in addition to increasing water use efficiency (WUE) and improving soil fertility. However, in spite of a significant increase in overall productivity, component crops suffer yield losses in intercropping systems due to competition for the limited divisible pool of growth resources. Therefore, there is an urgent need to optimize spatial and temporal arrangements in sorghum-legumes intercropping systems to realize maximum productivity and economic benefits (Iqbal et al., 2019).

9.3   LAND PREPARATION

The sustainability of agricultural ecosystems can be assessed by using soil enzyme activity as an indicator of soil quality. The effect of conservation tillage practices, for example, no tillage and reduced tillage (subsoil-bedding and shred-bedding), and conventional tillage practices, such as mold board plowing on physical-chemical, biochemical, and physical soil quality indicators was determined in a degraded sorghum field after a period of 3 years. A nearby soil representing local high quality soil and under native vegetation was utilized as a standard. The soil under conservation tillage, particularly no tillage had high crop residue accumulation. The tillage practices did not show any effect on soil electrical conductivity and pH. In the 0 to 5 cm layer, an increment in OM content was observed with the reduction in tillage intensity. The no tilled soil had 33% greater OM when compared with the average of the other tillage treatments. The higher values of water-soluble C, dehydrogenase, urease, protease, phosphatase, and β-glucosidase activities and aggregate stability were observed in no tilled soils. However, values were lower than the soil under native vegetation. They found that the enzyme activity and aggregate stability are more sensitive to soil management practices than physical-chemical properties. Thus, through their research findings they suggested that no tillage system could effectively improve the physical and biochemical qualities of soil (Roldána et al., 2005).

Dry land wheat (*Triticum aestivum* L.) and grain sorghum (*Sorghum bicolor* L.) are frequently cultivated using a wheat-sorghum-fallow (WSF) crop rotation on the semiarid North American Great Plains. The success of the WSF rotation depends on the precipitation stored during fallow as soil water. The evaporation can be reduced by stubble mulch-tillage (SM) and NT residue management practices. However, the sparse residue cover of dry land crops, like sorghum, is not enough to reduce soil crusting and runoff. Subsoil tillage practices such as paratill (PT) or sweep (ST), break infiltration reducing soil layers. These practices employed with residue management approaches, increases soil-water storage, and crop growth.

In this study the effects of PT to 0.35 m or ST to 0.10 m treatments on soil cone penetration resistance, soil-water storage, and dry land crop yield were compared with NT and SM residue management. Six contour-farmed level-terraced watersheds with a Pullman clay loam were cropped as pairs with a WSF rotation in such a way that each phase of the sequence occurred each year. In 1988, PT or ST was used on residue management plots during fallow after sorghum for every 3 years which resulted in five treatments: (i) NT-PT, (ii) NT-NOPT, (iii) NT-ST, (iv) SM-PT and (v) SM-NOPT. Highest cone penetration resistance was observed in NT plots and it decreased with PT after 12, 23, and 31 months. During fallow average soil water storage after wheat was more with NT than with SM. PT or ST did not show any effect on soil water storage. The consistent increase in dry land wheat and sorghum grain yields, total water use, and WUE were not recorded with NT. Their results specified that NT residue management is more useful for dry land crop production than subsoil tillage (Baumhardt and Jones, 2002).

In the semi-arid subtropics of central Queensland 10 years field trial was carried out in which 4 tillage practices (traditional, stubble mulch, reduced, no tillage) were compared under rainfed conditions on an alluvial soil. Further, responses to added nitrogen (N), sulfur (S), and zinc (Zn) fertilizers were ascertained in the final 4 years. To identify soil-conserving practices that can give high yields of high quality grain soil water storage, soil nitrate accumulation, grain yield (sorghum, wheat), grain protein content, and populations of soil macro fauna were measured. Under adequate soil fertility conditions, the yields in stubble mulch, reduced tillage, and no tillage plots were higher than traditional tillage used plots. The yield responses to tillage practices were owing to increases in soil water storage or crop WUE or both. The average soil macro fauna populations were 19 per m^2 in traditional tillage, 21 per m^2 in stubble mulch, 33 per m^2 in reduced tillage, and 44 per m^2 in no tillage. This influence of the tillage practices on soil animal populations might be a factor that contributes to the variations in soil water storage and WUE. Therefore, the study indicated that if adequate crop and fallow management practices are followed, conservation tillage practices could significantly improve grain yields (Radford et al., 1995).

9.4 WATER MANAGEMENT

Sorghum (*Sorghum bicolor*) is cultivated as a major summer crop in the subtropics. The high variations in rainfall and restricted planting chances make crop production risky in these regions. A strong crop simulation model can help farmers to measure production risks and make decision through simulation analyses. Therefore, a simple, still mechanistic crop simulation model was developed for sorghum, which assists in assessing climatic risk to sorghum production in water stress environments. The model simulates grain yield, accumulation of biomass, crop leaf area, phenology, and soil water balance. It makes use of a daily time-step and readily available weather and soil information and assumes no nutrient deficiency. The simulation model was examined on data (n=38) from experiments covering a diverse environmental conditions in the semi-arid tropics and subtropics. Probable limitations in the model were recognized and analyzed in a new testing procedure using combinations of projected and observed data in different modules of the model. The model performed satisfactorily, with 94% variation in total biomass and 64% variation in grain yield. These variations in outcome for biomass and yield were due to the limitations in calculating harvest index (Hammer and Muchow, 1994).

A field trial was carried out under continuous drying in the field to study the association between soil water extraction and root growth of grain sorghum. Soil water extraction was analyzed in terms of two factors: (1) the time when the extraction front reaches a depth, described as the moment when soil water content (θ) begins to decline exponentially with time and, (2) the decline of θ with time at each depth after the extraction front arrives. Further, the effect of critical assimilate limitation, affected by shading the crop during late vegetative growth on the rate of root front penetration and root length accumulation was assessed. The root front penetrated from sowing at a steady descending rate of 2.72 cm day^{-1}. The extraction front was behind the root front till 100 cm depth; afterwards they descended together, to get to the equal maximum depth of 190 cm just after anthesis. Total root length also attained a maximum at the same time. Shading did not show any influence on the rate of root front penetration and length accumulation, in spite of 70% reduction in aboveground growth. After the root front reached a particular depth, root proliferation continued till 40–20% of the extractable water in the layer remained. On account of the roots proliferating ahead of the extraction front, a little amount of water was extracted before the extraction front reached. The extraction rate at each depth attained a maximum when half of the extractable water remained. Therefore, the decline in θ can be better explained by a sigmoidal curve than an exponential curve, especially in the middle layers of the profile. However, the applicability of the sigmoidal curve to illustrate the decline in θ with time is limited (Robertson et al., 1993).

The effects of plant water stress at different stages of growth on grain and forage yields and yield components of grain sorghum (*Sorghum bicolor*) were studied. Irrigated grain sorghum was exposed to different periods of plant water stress starting at (i) early boot, (ii) heading, and (iii) grain filling. The yield of unstressed sorghum was 7,410 kg grain/ha. 13 to 15 days stress periods with mean afternoon leaf water potentials (LWP) –15.8 to –20.3 bars did not show any effect on grain yields. However, stress period of 27 to 28 days with mean afternoon LWP of 21.7 bars at early boot or heading stages cut down 27% yields. Conversely, a 27-day stress period (mean afternoon LWP, –22.7 bars) initiated at early grain filling lowered only 12% yields. Yield reductions of 43 and 54% were recorded when plants were subjected to 35 and 42-day stress periods (mean afternoon LWP, –22.9 and –24.0 bars) respectively at boot stage. Yield reductions from stress at early boot stage were results of reduced seed size and seed numbers. The stress imposed at heading or later, reduced only seed size. Stress periods of 27 to 28 days did not show any effect on forage yields. However, stress periods initiated at early boot stage and continued for 35 or 42 days reduced forage yields (Eck and Musick, 1979).

A field trail was undertaken to study the effect of *Azospirillum brasilense* inoculation on growth, water status, and yield of dryland sorghum (cv. RS 610 and cv. H-226). Plants were selected at regular intervals, and the characteristics such as dry-matter accumulation, leaf area, grain yield, percentage nitrogen and phosphorus in leaves, leaf water potential, canopy temperature, transpiration, stomatal conductance and soil water depletion were measured. The inoculated plants showed an average of 19% increase in total stover dry-matter yield due to high accumulation of dry-matter during the early stages of growth. Grain yield of *Azospirillum* inoculated plants was 15–18% higher than the non-inoculated plants in all three trails. This increment was correlated with a number of seeds per panicle. It was found that the inoculation improves water regime of sorghum plants as evident from their higher leaf water potential, lower canopy temperatures and greater stomatal conductance and transpiration. The inoculated plants extracted water deeper soil layers and had 15% greater extraction of soil moisture when compared with non-inoculated controls. These results indicated that inoculation with *Azospirillum* can increase yield of dry land grain sorghum, mainly by improving exploitation of soil moisture (Sarig et al., 1988).

In the central and southern high plains of the USA, groundwater is usually extracted for irrigation. Benefits and risks in irrigation management are influenced by the relationship between crop yield and level of water applied. A 14-year period (1974–1987) study was undertaken to investigate the association between yield and water application in corn, grain sorghum, and sunflower on silt loam soil. Plots were level-basins and water was applied separately through gated pipe. With the increase in amount of irrigation from 100 to 200, 200 to 300, and 300 to 400 mm, sunflower yield increased by 0.53 Mg ha^{-1}, 0.43 Mg ha^{-1}, and 0.37 Mg ha^{-1}, respectively. Corn yield was higher than grain sorghum at total irrigation amounts of 345 mm and above. Yield increase in corn was greater than in grain sorghum over continuous dry land at total irrigation amounts above 206 mm. Thus, if grain mass is considered, grain sorghum is a better option than corn at less than 206 mm of irrigation, however at more than 206 mm of irrigation corn is a better alternative than grain sorghum (Gwin and Khan, 1996).

To make conjunctive use of rainfall and limited irrigation of graded furrows a limited irrigation-dry land (LID) farming system was developed and tested under field conditions. The LID system utilizes a less water supply to irrigate a larger area than could be conventionally irrigated. The distinctive feature of this design is that it actually adjusts, the amount of land irrigated during the crop growing season, as higher area can be irrigated during above-average rainfall years with less amount of irrigation water than during below average years. A 600 m long graded furrow field, on a 0.3 to 0.4% slope was divided into three water management sections. The upper half of the field was "fully irrigated." The following one fourth was maintained as a "tail water runoff" section that makes use of furrow runoff from the fully irrigated section. The lower one-fourth was maintained as a "dry land" section that can use runoff from either rainfall or irrigation on the wetter sections. The LID system was tested under field conditions for 3 years by cultivating grain sorghum (*Sorghum bicolor*) on a Pullman clay loam (fine, mixed, thermic family of *Torretic paleustolls*). When compared to the average rainfall of 250 mm, seasonal rainfall in 1979, 1980, and 1981 were 224 mm, 103 mm and 424 mm, respectively. The LID system successfully increased the utilization efficiency of irrigation water. After harvest, irrigation water applied to a field can be lost through transpiration, evaporation, percolation or as runoff, or left stored in the soil. The LID system has a tendency to decrease all losses other than transpiration. When grain sorghum was applied with irrigation water of 125 mm or 185 mm with the LID system, the evapotranspiration was increased by an almost equal amount. This increased an average grain yield of 154 kg/ha for each 10 mm added irrigation, compared to 92 kg/ha for each 10 mm under conventional irrigation practices (Stewar et al., 1983).

9.5 FERTILIZER MANAGEMENT

Nitrogen fertilizer is a costly and basic input for optimal production of non-leguminous crops. The double cropping of summer grain crops with leguminous winter crops can reduce the amount of fertilizer applied to grain crops. However, costs of seeding winter legumes are often as costly as the commercially produced N that they substitute. The cost of legumes seeding can be eliminated by permitting the legumes to produce seed before planting the summer crop. Therefore, to determine the requirements of N fertilizer for NT grain sorghum double cropped with reseeding crimson clover and effects of N treatments on nutrient uptake and on insect populations an experiment was carried out on a Cecil sandy loam soil. The nitrogen doses 0, 15, 30, 45, 90, and 135 kg N/ha was included as treatments and the clover tissue was either taken away or left on the soil surface as a NT mulch. The planting of No-till sorghum into clover or removal of clover top growth at maturity did not show noticeable effect on clover stand establishment. Significant difference was not observed in clover DM production among years and the average production was 4,762 kg/ha/yr. The average concentrations of N, P, and K in the clover tops were 2.02, 0.24, and 2.40%, respectively. The Cu content in the sorghum leaves increased with the application of N at the bloom stage. The reduction in K in 1979 (from 2.47 to 2.31%), N in 1980 (from 2.85 to 2.69%), and P in 1979 (from 0.36 to 0.34%) and 1980 (from 0.34 to 0.29%) of sorghum leaves was observed when clover was removed. However, these reductions were not enough to produce a deficiency of either element. Some variances were observed in common insect populations and their successive damage among treatments. However, the insect's populations were not above economic threshold level. Nitrogen supplied by the clover was adequate for highest sorghum grain yield without additional applications of inorganic N. The removal of clover did not show any influence on sorghum grain yield in 1978 or 1979, but in 1980 this treatment decreased the grain yield. Although removal of clover containing 62 kg/ha N in 1980 decreased sorghum yield, there was no response to applied N which indicates that the reduction in yield was not because of N shortage (Touchton et al., 1981).

An investigation on the effects of nitrogen deficiency on photosynthetic rates and hyper reflectance properties of sorghum by Dulizhao et al. (2005) has shown that a decline in the N contents in the Hoagland solutions when applied to the plants grown under greenhouse conditions led to a decline in the leaf chlorophyll contents, leaf area and photosynthetic rates in the plants under nitrogen deficiency with a simultaneous reduction even in the production of biomass. Under nitrogen deficiency the plants exhibited lower photosynthetic rates due to reduced stomatal conductance than to a decline in the carboxylation capacity of the leaves. Similarly, N deficiency resulted in a drastic decline in the leaf dry weights than that of the roots. Though the leaf reflectance exhibited an increase under nitrogen deficit conditions at a longer wavelength (715 nm) a red edge shift was much towards the shorter wavelength. Further, their leaf N and Chl concentrations were linearly correlated with both reflectance ratios of R_{405}/R_{715} ($r^2 = 0.68^{***}$) and R_{1075}/R_{735} ($r^2 = 0.64^{***}$), respectively as well as first derivatives of the reflectance ($dR/d\lambda$) in red edge centered

730 or 740 nm (r^2 = 0.73–0.82***). Therefore, rapid and non-destructive estimation of sorghum leaf Chl and plant N status can be made using these specific reflectance ratios or $dR/d\lambda$.

The application of mineral fertilizer in small amounts to the planting hole or pocket known as micro fertilizing was analyzed in Mali. Application of 0.3 g fertilizer per pocket increased sorghum yield by 34% and 52% when compared with the control during 2000 and 2001, respectively. With the increase in fertilizer dosage from 0.3 g to 6 g corresponding increase in sorghum grain yield was observed. The value cost ratio varied from 3.4 to 11.9 in the 0.3 g treatment, and from 0.43 to 1.17 in the 6 g treatment. Therefore, owing to its good return on investment, low financial risk, low cost and low workload 0.3 g of micro fertilizer application can be recommended to farmers (Jens et al., 2007).

An experiment was carried out under fully irrigated conditions in tropical Australia to assess the effect of variable nitrogen supply on leaf initiation, expansion, and senescence and on leaf nitrogen in maize and sorghum. It was found that the development of leaf area in sorghum was more responsive to N than in maize. Highest leaf area indices were noted for sorghum and found to be related with long vegetative development period and high plant population density of sorghum. The primary effect of variable nitrogen supply was observed on leaf expansion, with variations in leaf initiation being comparatively small. The minimum amount of leaf N per unit leaf area needed for leaf expansion termed as minimum specific leaf nitrogen (SLN) was proposed. The data indicated that if N uptake and distribution to leaves is adequate to retain SLN above 1.0 g m^{-2} then leaf expansion progresses at its potential rate. Little leaf senescence was observed during the vegetative period of sorghum at low rates of applied N, however during grain filling it was more prominent. Further, a marked reduction in SLN was noticed during the grain filling stage of sorghum, at lower rates of applied N (Muchow, 1988).

Sorghum is well adapted to African climate patterns and is expected to sustain extensive suitability across different African climatic zones under climate change. Under a changing climate, drought tolerance of sorghum and its capacity to withstand waterlogging make it a significant crop to maintain productive agro-ecosystems. As sorghum is a staple grain in Africa, improved sorghum management can provide stability in the smallholder farmer's household nutritional needs. In this chapter, meta-analysis was used to review sorghum yield trends across nutrient management setups. Under eight nutrient management practices treatments viz. N-only, P-only, N, and P, N and P microdose, legume management, manure addition, OM amendment, and mixed amendment sorghum yields were compared. When compared with control 47–98% higher yield was recorded under chemical fertilizer amendment. The organic nutrient amendment increased grain yield by 43–87%. However, the magnitude of yield increase is not the only factor that determines the profitability of a management scenario. For example, though application of nitrogen (N) and phosphorus (P) produced the highest yield, the profit was lowest owing to the high cost of fertilizer. On the other hand, the rotation of sorghum with an edible legume increased 43% yield when compared to no nutrient inputs with a net profit of US $146 to $263 per hectare. Therefore, easy accessibility to both fertilizer inputs and varied rotations has the highest potential to improve sorghum grain yield in Africa (Christina and Jacob, 2016).

Nitrogen (N) is a vital nutrient for crop growth and development but its low availability and loss is a foremost limiting factor in crop production. An experiment was carried out to study the effects of N fertilization on WUE and physiological and yield traits of sorghum. Three sorghum varieties were assessed by applying six N-levels (0, 20, 40, 60, 80, and 100 kg ha^{-1}) at a constant phosphorus and potassium level of 30 kg ha^{-1}. Findings revealed that N increased grain yield by 35–64% at the Bayero University Kano (BUK) and 23–78% at Minjibir. The maximum average grain yield was recorded at the N-fertilizer treatment at 80 kg N ha^{-1}. Among the N-levels and among the sorghum varieties significant differences were noticed for assessed WUE. N-fertilizer treatments increased WUE by 48–55% at BUK and by 54–76% at Minjibir over control treatment. Maturity and physiological trait showed a significant influence on WUE. The highest average grain yield and WUE were recorded in the extra early maturing variety (ICSV400), whereas lowest WUE was recorded in late maturing variety (CSR01) (Hakeem et al., 2018).

As winter legume cover crops conserve soil and water resources and may provide significant amounts of N, the role of these legumes in conservation tillage production systems needs improved attention. This experiment was conducted to ascertain the amount of N contributed by legumes to a succeeding grain sorghum crop under no tillage and the effect of legume cover crops on soil fertility status. A field trail was conducted for three years using four winter legumes, one non-legume and a no cover crop as treatments. N was applied at four rates, i.e., 0, 28, 56, and 112 kg N ha^{-1} to no-till grain sorghum. The grain sorghum grown in succession with a legume cover crop did not show any response to N fertilizer. However, responded to as much as 99 kg N ha^{-1} when succeeding a non-legume cover crop or no cover crop. It was found that legume replaces about 72 kg N ha^{-1} obtained from the fertilizer N. Some of effects of legume cover crops on soil fertility are reduction in pH, reallocation of K$^+$ to the soil surface from deeper layers in the soil profile, and lower C/N ratio in soil OM. As N fertilizer provides a large portion of the fossil fuel energy needed for non-leguminous row crop production, the projected N contribution of legume cover crops denotes a significant energy savings, which improves the conservation value of a NT production system (Hargrove, 1986).

In comparison to fresh manure, composting lowers material quantity, hauling, and spreading costs, odor, and weed seed viability. Further, it improves physical properties of soil. However, the amount of nutrients supplied by manure compost may be less. In this study, beef cattle manure compost was evaluated as a source of nutrient for irrigated grain sorghum and the influences of annual compost and N fertilizer treatment son soil chemical properties were determined. In a factorial design, manure compost was applied at five rates, i.e., 0, 0.9, 1.8, 3.6, and 7.2 ton DM/acre and N fertilizer at four rates, i.e., 0, 55, 110, and 165 lb/acre on a Ulysses silt loam (fine silty, mixed, mesic, Aridic Haplustoll) at Tribune, KS. The plots applied with combinations of compost and N fertilizer recorded higher grain yields than from either applied alone. The evident fertilizer efficiency for N fertilizer was 36%, while for compost it was 13%. An increment in soil P, K, and OM were observed with increasing rates of compost. On the other hand, increased rates of N fertilizer reduced soil P and K, without any effect on OM. The addition of compost increased soil Na levels but not to extreme levels. The application of compost did not show any effect on soil NO$_3$-N levels, but nitrogen fertilizer increased NO$_3$-N levels. Therefore, study results suggested that manure compost maintains or increases soil chemical levels, particularly P, without excessive accumulation of NO$_3^-$; however, it should be applied with N fertilizer to improve grain production (Schlegel, 1992).

To assess the residual effects of soybean on N uptake, agronomic characters and yield of the sorghum crop and residual soil N accumulation, a two years field trial was conducted by growing soybean (*Glycine max*) and grain sorghum (*Sorghum bicolor*) in rotation. Both the continuous sorghum and sorghum grown in rotation were applied with N fertilizers at the rates of 0, 56, 112, and 168 kg N ha^{-1}. The crop rotation significantly increased the total N uptake by sorghum at the fertilizer rate of 112 kg ha^{-1} or less. Sorghum cultivated in rotation with soybean and applied with 56 kg N ha^{-1} or more flowered 6.1 days before than continuous sorghum without N. At the zero N fertilizer level, the grain yield of sorghum grown in rotation was 85% more than the continuous sorghum. At or above the 112 kg N ha^{-1} level, significant variances were not observed between the cropping systems. When 56 kg N ha^{-1} or less was applied, the residual soil N as NO_3^-N after soybean was 50 to 60 kg N ha^{-1} more than after continuous sorghum. Thus, their studies have indicated that 56 kg N ha^{-1} or less is the optimum dose of N for sorghum grown in rotation with soybean and it has positive effects on sorghum grain yield (Gakale and Clegg, 1987).

Sweet and fiber sorghum (*Sorghum bicolor*) are multi-use cereals of primary interest. During 1997–1999 field trials were conducted to evaluate the effects of nitrogen (N) fertilization on crop growth, yield, and N budget throughout crop growth period. Sweet and fiber sorghum were applied with N at three rates, i.e., 0, 60, 120 kg ha^{-1}. A sigmoidal growth-pattern was shown by both the crops, but an earlier and steeper growth was observed in fiber sorghum. Nitrogen fertilization had insignificant effect on growth pattern or yield partitioning among plant organs. The sweet sorghum shared more photosynthates to stems viz. around 75% of total dry weight than fiber sorghum (55–60%), owing to a reduced partitioning to panicles. An interaction was found between total dry weight at harvest and years: fiber sorghum gave significantly higher yields than sweet sorghum in 1997 (23.8 versus 17.7 mg ha^{-1}), however the opposite occurred in 1998 and 1999 (average 20 versus 24.2 mg ha^{-1}), when another sweet hybrid of longer season was grown. When only the stem was taken into account of potential interest, such as in the case of ethanol production or fiber extraction, fiber sorghum exhibited some advantage in the 1st year (14.4 versus 12 mg ha^{-1}), while the sweet type predominated in the next 2 years (18.7 versus 11 mg ha^{-1}). Fertilization had no significant effect on total yield, though it showed an interaction with sorghum type in plant N concentration and uptake. Higher N concentration and uptake was observed in fiber sorghum because of its larger panicles which acts as a late-season sink for the nutrient. Applied fertilizer and plant uptake exhibited a strong effect on nitrogen budget at harvest, while nutrient losses as ammonia volatilization and nitrate leaching and natural provisions with precipitation enacted a less relevant role. Depending on the plant portion removed from the field, the predictable variations in soil reserves ranged between −216 and +77 kg ha⁻¹ of N. An insignificant reduction in soil resources, more favorable in environmental terms, was attained when the whole biomass was taken away from the field and when fertilizer rates were adequate to crop requirements under particular growth conditions (Lorenzo et al., 2005).

In the humid tropics, the major constraint in sorghum production is leaching loss of N on permeable soils and under heavy rainfalls. An experiment was conducted in the central Amazon (near Manaus, Brazil) to analyze the effect of charcoal and compost on the N retention. After fifteen months of organic-matter admixing (0–0.1 m soil depth), 15N-labeled $(NH_4)_2 SO_4$ (27.5 kg N ha⁻¹ at 10 atom % excess) was added. The tracer was evaluated in top soil (0–0.1 m) and plant samples collected at two continuous sorghum harvests. As a result of considerably higher crop production during the first growth period, the plots applied with compost (14.7% of applied N) had significantly higher N recovery in biomass when compared with the mineral-fertilized plots (5.7%). After the second harvest, the charcoal-amended plots (15.6%) showed significantly higher N retention in soil than only mineral-fertilized plots (9.7%). The retention of applied N fertilizer was increased by the organic amendments. In compost (16.5%), charcoal (18.1%), and charcoal-plus-compost treatments (17.4%) significantly higher total N recovery in soil, crop residues, and grain yields were recorded when compared with mineral-fertilized treatment (10.9%). It was found that the recycling of N is one of the processes taken up by the crop in this retention. Therefore, the importance of immobilization, decreased N leaching, and gaseous losses together with other potential methods for improving N retention should be resolved in forthcoming studies (Wenceslau et al., 2008).

Grain sorghum (*Sorghum bicolor*) was cultivated in rotation with wheat (*Triticum aestivum* L.) and soybean (*Glycine max*) in the mid-Atlantic. The influence of N fertilization at multiple rates on dry land sorghum was evaluated. Four starter-band N rates (11, 34, 56, and N ha⁻¹) and four side-dress N rates (0, 45, 90, and 134 kg N ha⁻¹) were applied as treatments in factorial experimental design. A broadcast treatment of 67 kg N ha⁻¹ at planting was also incorporated. Starter-band was applied 5 cm to the side and below the seed. At the eight-leaf growth stage viz., 35 days after emergence side dress was applied. The average grain yield over eight years ranged from 1.7 to 11.9 Mg ha⁻¹. Yield was responsive and nonresponsive to N applications on four sites each. Either high levels (>85 kg N ha⁻¹) of residual soil mineral N, or acute water stress conditions were responsible for non-responsiveness. The study findings indicated that the combination of starter-band N and side-dress N (130 kg N ha⁻¹) should not be applied to the soils high in mineral N (50 kg N ha⁻¹ in the surface 0.3 m) for most favorable economic return to N fertilization. However, the application of 40 kg N ha⁻¹ starter-band in combination with 130 kg N ha⁻¹ side-dress N can improve the sorghum yields in soils with low mineral N (Raj et al., 2000).

Forage sorghum-Sudan grass (*Sorghum bicolor* (L.) Moench) is a quite new crop to eastern Canada. The influences of fertilizer N supplement son yield, N accumulation, and N use efficiency of this crop were studied in this region. The response of forage sorghum-Sudan grass to fertilizer N applied at different rates (0, 50, 100, 150, 200, and 250 kg N ha⁻¹) either as a single side dress or split into two-side dress was assessed on a Fox loamy sand. However, timing of N application had insignificant effect on DM production; the application of N in two equal splits enhanced the NUE and ANR of individual cuts. Highest grain yield (5.95 Mg ha⁻¹) was recorded at N rate of 125 kg N ha⁻¹. Depending on the cost of N fertilizer and value of hay, the most economic rates of N ranged from 83 to 107 kg N ha⁻¹. An increment in nitrogen concentration was observed with increasing N application and the highest N accumulation was 161 kg N ha⁻¹ at N rate of 196 kg N ha⁻¹. However, total N use efficiency and apparent N recovery get reduced with increasing N rates varying from 36 to 11 kg DM kg⁻¹ and 90 to 24%, respectively. Optimal yield and N efficiency were recorded when 100 kg N ha⁻¹ was applied in two splits. Therefore, the producers in southern Ontario need N fertilizer supplements to improve sorghum-Sudan grass yields but they should avoid over-fertilization with N to increase N use efficiency and apparent N recovery (Ronald and Robert, 2005).

A greenhouse study was conducted in which pelletized compost was used as a soil amendment and plant nutrient source after incorporation in Arredondo sand. The application of 8 metric tons/ha of compost crops increased the yields of two sorghum crops (*Sorghum bicolor* L. Moench). Higher yields were recorded at highest rate of compost (64 metric tons/ha) application as compared to application of 10-4.4-8.3 fertilizer at 2 tons/ha. Compost applications increased the uptake of all plant nutrients measured, except Mn. Furthermore, compost applications improved the water retention and cation exchange capacity of the Arredondo sand (Hortenstine and Rothwell, 1973).

Effective use of external inputs and water conservation are essential for sustainable agricultural productivity in semiarid West Africa. N mining in poor fertile soils of West Africa can be lessened through an incorporation of local water and nutrient management practice was concluded by field experimental results conducted for three years by Robert et al. (2004), on a Ferric Lixisol with 1% slope. They assessed the effects of stone rows or grass strips of *Andropogon gayanus* Kunth cv. *Bisquamulatus* (Hochst. Hack.) on soil and water conservation (SWC) measures. The SWC measures were placed along contours lines. They detected that during the three consecutive years, all treatments induced negative annual N balances (-75 to -24 kg N ha^{-1}). N transfers by sorghum biomass and soil erosion-induced N losses were key factors elucidating these negative balances. Runoff and eroded sediments caused large amounts of N (7 kg N ha^{-1} per year in 2000 and 44 kg N ha^{-1} per year in 2002) loses in the control treatment, which corresponds, respectively, to about 10 and 43% of the total outflow of N. Erosion N losses were reduced to 8 and 12% of the total annual loss by sole stone rows and grass strips, respectively. The combined application of SWC measures and nutrients inputs decreased erosion N losses to only 2–7% of the annual N loss. The lowest soil N mining was observed in plots applied with urea-N or compost-N over the 3 years, whereas the plots without added N had highest N mining.

Grain sorghum [*Sorghum bicolor* (L.) Moench] is cultivated in rotation with wheat (*Triticum aestivum* L.) and soybean [*Glycine max* (L.) Merr.] in the mid-Atlantic. Adequate data on N fertilization of sorghum are not available for this region. Raj et al. (2000) in their evaluation of the effect of multi-rate N fertilization on dry land sorghum found that either high levels (>85 kg N ha^{-1}) of residual soil mineral N, or severe water stress conditions was responsible for non-responsiveness. The study findings indicated that when sorghum is grown on soils with high mineral N (50 kg N ha^{-1} in the surface 0.3 m)starter-band N in combination with side dress N of 130 kg N ha^{-1} should not be applied at planting for optimum economic return to N fertilization. However, in soils with low mineral N, 40 kg N ha^{-1} application of starter-band in combination with 130 kg N ha^{-1} side-dress N could improve the sorghum yields in most situations.

9.6   RESEARCH ADVANCES IN CULTIVATION OF SORGHUM

A study has been undertaken to measure transpiration efficiency (W), the ratio of plant carbon yielded to water transpired and carbon isotope discrimination of leaf DM, Δd, on 30 sorghum lines in the glasshouse and on 8 lines grown in the field. In the glasshouse, mean W was 4.9 m mol C mol^{-1} H$_2$O and mean Δd was 3.0. In the field, the mean W observed was 2.8 m mol C mol^{-1} H$_2$O and the mean Δd was 4.6%. The plants grown in both greenhouse and field showed positive and significant correlations between W and Δd. Further, gas exchange of carbon isotope discrimination during CO$_2$ uptake, ΔA, and pi/pa were measured in detail on leaves of eight sorghum lines. A linear and negative association was observed between ΔA and pi/pa. ΔA showed a negative slope of 3.7% for a unit change in pi/pa and this slope was found be reliant on the leakiness of the CO$_2$ concentrating mechanism of the C$_4$ pathway. The leakiness was estimated to be 0.2. The study concluded that the variation in Δd reflects the variation in transpiration efficiency, W, though observed variation in the 30 lines of sorghum is smaller than that usually seen in C$_3$ species (Henderson et al., 1998).

As stem length and diameter determines stem volume and yield, when thinning and weeding was practiced up to 1 month after sowing plants with heavy-stem and high yield were produced. Stem yield of 4,790–6,593 g m^{-2} was recorded in Wray cultivar. Stem sugar weight increased after anthesis and maximum sugar weight was recorded at 26–30 days after anthesis (DAA). Grain grew up to 26 DAA and dried. Grains attained harvestable maturity after 30 DAA. Strong correlation was observed between stem sugar content and Brix, which indicated that Brix value is an index of sugar in stem juice. Therefore, the optimum harvest time was ascertained by measuring Brix after 30 DAA (103 DAS). Maximum stem and sugar yields were recorded in cv. Wray. Grain yield of other cultivars was comparable to cv. Wray's. Utilization of single-step milling without imbibition water provided a juice extraction rate of 50%. However, multiple milling steps with imbibition water yielded more than 400 g m^{-2} of sugar with 75% juice extraction rate. Thus the results suggested that sweet sorghum could be cultivated on dry land in the rainy season (Naoyuki and Yusuke, 2004).

In most of the agrosystems Arbuscular mycorrhizal fungi (AMF) constitute major component of soil microbiota. The AMF colonize the roots of the plants as obligate mutualistic symbionts. Here, the effect of different tillage systems (moldboard, shred-bedding, subsoil-bedding, and no tillage) on the AM fungal community colonizing maize, bean, and sorghum roots was studied using molecular techniques. Roots from 36 plants were examined with AM fungal-specific primers to amplify the small subunit (SSU) of the ribosomal DNA (rDNA) genes partially. About 880 clones were screened for restriction fragment length polymorphism (RFLP) variation, and 173 clones were sequenced. Ten AM fungal types were detected and grouped into three AM fungal families, i.e., Gigasporaceae, Glomaceae, and Paraglomaceae. In all the samples the predominant taxon was glomus. Out of these ten AM fungal types four types were different from any earlier published sequences and could relate to either known unsequenced species or unknown species. Low fungal diversity was found in the four agriculture management methods. However, the multidimensional scaling (MDS) analysis and log-linear-saturated model revealed that the tillage system significantly effects the composition of AMF community. To conclude, as some fungal types were specific to treatment, agricultural practices could directly or indirectly effect biodiversity of AM (Lumini et al., 2008).

In semi-arid areas variability of rainfall has a significant effect on crop production with manifestations in recurrent crop failure. Here, parameterized APSIM crop model was used to study the effect of rainfall irregularity on yields of improved sorghum varieties. The long-term historical rainfall and anticipated climate was analyzed, which revealed non-significant and significant trends on the onset, cessation, and extent of the growing season. Based on seasonal rainfall distribution, simulated sorghum yields were analyzed, that indicated the correspondence of lower grain yields with the 10-day dry spells during the cropping season. However, the results of simulation for future sorghum response showed that the effects of rainfall irregularity on sorghum will be reversed by increasing temperature. Therefore, the study proved that variability in rainfall indeed influences yields of improved sorghum varieties. However, in the event where harms imposed by moisture stress are not halted, even improved sorghum varieties are probable to perform poorly (Barnabas et al., 2017).

Land suitability evaluation helps in recognizing the inherent land potentials and limitations through which the difference between the actual requirements and actually applied in a given land could be avoided. In this study geographic information system (GIS), fuzzy set models and analytical hierarchy process (AHP) methods were integrated to assess the land suitability for sorghum crop. Characteristics such as soil, climate, and topography were taken into consideration. In northern semi-arid Ethiopia the model results revealed that 29,534 ha (30.54%) of the area was moderately suitable, 34,984.74 ha (36.17%) of the area was marginally suitable, 17,455 ha (18.05%) of the area was currently not suitable and 14,744.61 ha (15.24%) of the area was permanently not suitable for sorghum crop production. The foremost limiting factors for sorghum production in the area were slope gradient, altitude, temperature, length of growing period, available water capacity, average weight diameter, total nitrogen, available phosphorus, and soil organic carbon contents. The study proposed that the adequate organic and inorganic fertilizer application, tillage, and soil and water management practices are required to increase sorghum productivity in the area (Araya et al., 2018).

Sweet sorghum prefers warm moist soil for germination and emergence. However, it would be more beneficial if it can be grown in various seasons. As the crop uses less water and has a short life cycle when compared with sugarcane, the probability of growing the crop throughout the year was assessed. Sixteen sweet sorghum genotypes collected from different areas of Sudan were evaluated under winter conditions. Among genotypes significant differences were noted for days to germination, plant height, number of leaves per plant, chlorophyll content, stem diameter, head weight, shoot fresh weight, head to shoot ratio, brix value, juice weight and number of days to maturity. They observed that there was strong positive association between juice and shoot weight and negative association between brix value and head weight. The genotypes exhibited high variability in all parameters, therefore, could be valuable genetic resources for breeding winter adaptation (Mohammed et al., 2019).

The dynamics of food production should be understood to improve food security. Especially in areas that depend on subsistence agriculture with less adaptive capacity to climate change. In some of the poorest sections of the world sorghum plays a pivotal role in food security. Here, the literature was reviewed to find and study the key factors influencing sorghum production in three major production regions North America, Asia, and Africa's Sahel. It was found that climate change, population growth/economic development, demand for non-food, agricultural inputs, other crops demand, shortage of agricultural resources, biodiversity, cultural influence, price, and armed conflict were major factors influencing sorghum production. The study revealed that the sorghum production is affected by multiple factors simultaneously. The extent and certainty of one or more other factors determines the effect of each factor. In different geographical regions factors vary in relevance and degree. In general, improved agricultural inputs, population growth/economic development and climate change have significant effect on production of sorghum. However, local dynamic forces expected to go beyond these wide-ranging trends and more in-depth, locally concentrated studies are required for actionable planning purposes (Clara et al., 2019).

9.7   CONCLUSIONS

Various phases of sorghum cultivation are presented in this chapter. These comprised land preparation, cropping systems, water management, fertilizer management, harvesting storage and marketing. In addition research advances in sorghum cultivation are discussed.

KEYWORDS

- cultivation
- harvesting
- irrigation
- marketing
- physiological maturity
- sorghum

REFERENCES

Andrew, O. E., Steven, N. N., & Catherine, W. M., (2016). Effect of facilitative interaction of sorghum-cowpea intercrop on sorghum growth rate and yields. *Journal of Environmental and Agricultural Sciences, 9,* 50–58.

Araya, K., Mitiku, H., Girmay, G., & Muktar, M., (2018). Land suitability analysis for sorghum crop production in northern semi-arid Ethiopia: Application of GIS-based fuzzy AHP approach. *Cogent Food and Agriculture, 4,* 1–24.

Barnabas, M. M., Tumbo, S. D., Kihupi, N. I., & Filbert, B. R., (2017). Performance of sorghum varieties under variable rainfall in Central Tanzania. *International Scholarly Research Notices*, 1–10.

Baumhardt, R. L., & Jones, O. R., (2002). Residue management and tillage effects on soil-water storage and grain yield of dry land wheat and sorghum for a clay loam in Texas. *Soil and Tillage Research, 68*, 71–82.

Christina, T., & Ricker-Gilbert, J. E., (2016). Nutrient management in African sorghum cropping systems: Applying meta-analysis to assess yield and profitability. *Agronomy for Sustainable Development, 36*, 10.

Clara, W. M., Silvia, S., Kofi, A., & Guangxing, W., (2019). A regional comparison of factors affecting global sorghum production: The case of North America, Asia, and Africa's Sahel. *Sustainability, 11*, 1–18.

Duli, Z., Raja, R. K., Vijaya, G. K., & Reddy, V. R., (2005). Nitrogen deficiency effects on plant growth, leaf photosynthesis, and hyperspectral reflectance properties of sorghum. *European Journal of Agronomy, 22*, 391–403.

Eck, H. V., & Musick, J. T., (1979). Plant water stress effects on irrigated grain sorghum: I. effects on yield. *Crop Science, 23*, 81–90.

Gakale, P., & Clegg, M. D., (1987). Nitrogen from soybean for dry land sorghum. *Agronomy Journal, 79*, 1057–1061.

Ghosh, P. K., Ajay, K., Bandyopadhyay, K., Manna, K., Mandal, K., Misra, A., & Hati, K. K., (2004). Comparative effectiveness of cattle manure, poultry manure, phosphocompost and fertilizer- NPK on three cropping systems in vertisols of semi-arid tropics: II. Dry matter yield, nodulation, chlorophyll content and enzyme activity. *Bioresource Technology, 95*, 85–93.

Gwin, R. E., & Khan, A. H., (1996). Response of corn, grain sorghum, and sunflower to irrigation in the high plains of Kansas. *Agricultural Water Management, 30*, 251–259.

Hakeem, A. A., Folorunso, M. A., Kunihya, A., & Jerome, J., (2018). Productivity and water use efficiency of sorghum (*Sorghum bicolor*) grown under different nitrogen applications in Sudan savanna Zone, Nigeria. *International Journal of Agronomy*, 1–11.

Hammer, G. L., & Muchow, R. C., (1994). Assessing climatic risk to sorghum production in water-limited subtropical environments, development, and testing of a simulation model. *Field Crops Research, 36*, 221–234.

Hargrove, W. L., (1986). Winter legumes as a nitrogen source for no-till grain sorghum. *Agronomy Journal, 78*, 70–74.

Henderson, S. S., Von, C. G. D., Farquhar, L., & Hammer, G., (1998). Correlation between carbon isotope discrimination and transpiration efficiency in lines of the C_4 species *Sorghum bicolor* in the glasshouse and the field. *Australian Journal of Plant Physiology, 25*, 111–123.

Hortenstein, C. C., & Rothwell, D. F., (1973). Pelletized municipal refuse compost as a soil amendment and nutrient source for sorghum. *Journal of Environmental Quality, 2*, 343–345.

Iqbal, M., Muhammad, A. H., Abdul, A., Tanvir, S., Muzammil, H., Imtiaz, A., Sajid, A., Anser, A., & Zahoor, (2019). Forage sorghum-legumes intercropping: Effect on growth, yields, nutritional quality, and economic returns. *Bragantia, 8*, 78–85.

Jens, B. A., Mamadou, D., & Abou, B., (2007). Micro fertilizing sorghum and pearl millet in Mali. Agronomic, Economic, and Social Feasibility. *Outlook on Agriculture, 36,* 199–203.

Lorenzo, B., Silvia, G., Angela, V., & Gianpietro, V., (2005). Sweet and fiber sorghum (*Sorghum bicolor* (L.) Moench), energy crops in the frame of environmental protection from excessive nitrogen loads. *European Journal of Agronomy, 23,* 30–39.

Lumini, E., Roldán, A., & Salinas-García, J. R., (2008). The impact of tillage practices on arbuscular mycorrhizal fungal diversity in subtropical crops. *Ecological Application, 18,* 527–536.

Mohammed, A. E. B., Ghazi, H. B., Alfred, O., & Lembe, S. M., (2019). Assessment of the effects of winter condition on sweet sorghum yield and sugar content. *Turkish Journal of Agriculture-Food Science and Technology, 7,* 166–172.

Muchow, R. C., (1988). Effect of nitrogen supply on the comparative productivity of maize and sorghum in a semi-arid tropical environment, leaf growth, and leaf nitrogen. *Field Crops Research, 18,* 1–16.

Naoyuki, T., & Yusuke, G., (2004). Cultivation of sweet sorghum (*Sorghum bicolor* (L.) Moench) and determination of its harvest time to make use as the raw material for fermentation, practiced during rainy season in dry land of Indonesia. *Plant Production Science, 7,* 4–10.

Nguyen, D. C., & Tomohiko, Y., (1999). Genotypic and phenotypic variances and covariances in early maturing grain sorghum in a double cropping. *Plant Production Science, 6,* 89–95.

Paul, W. U., & Allen, F. W., (1979). Managing irrigated winter wheat residues for water storage and subsequent dry land grain sorghum production. *Crop Science, 9,* 89–95.

Radford, B. J., Key, A. J., Robertson, L. N., & Thomas, G. A., (1995). Conservation tillage increases in soil water storage, soil animal populations, grain yield, and response to fertilizer in the semi-arid subtropics. *Australian Journal of Experimental Agriculture, 35,* 223–232.

Raj, K., Mark, M. A., & Paul, H. D., (2000). Nitrogen management in no-tillage grain sorghum production: I. Rate and time of application. *Agronomy Journal, 92,* 23–29.

Refay, Y. A., Alderfasi, A. A., Selim, M. M., & Awad, K., (2013). Growth and yield production of grain sorghum-cowpea under limited water supply condition growth, yield and yield component characters of sorghum. *IOSR Journal of Agriculture and Veterinary Science, 2,* 24–29.

Robert, Z. B., Mando, A., Stroosnijder, L., & Guillobez, S., (2004). Nitrogen flows and balances as affected by water and nutrient management in a sorghum cropping system of semiarid Burkina Faso. *Field Crops Research, 90,* 235–244.

Robertson, M. J., Fukai, S., Ludlow, M. M., & Hammer, C., (1993). Water extraction by grain sorghum in a sub humid environment: II. Extraction in relation to root growth. *Field Crops Research, 33,* 99–112.

Roldána, A., Salinas-Garcíab, J. R., Alguacila, M. M., Díazc, E., & Caravacaa, F., (2005). Soil enzyme activities suggest advantages of conservation tillage practices in sorghum cultivation under subtropical conditions. *Geoderma, 129,* 178–185.

Ronald, P. B., & Robert, C. R., (2005). Influence of nitrogen fertilization on multi-cut forage sorghum-Sudan grass yield and nitrogen use. *Agronomy Journal, 97,* 45–56.

Sarig, S., Blum, A., & Okon, Y., (1988). Improvement of the water status and yield of field-grown grain sorghum (*Sorghum bicolor*) by inoculation with *Azospirillum brasilense*. *The Journal of Agricultural Science, 110*, 271–277.

Schlegel, A. J., (1992). Effect of composted manure on soil chemical properties and nitrogen use by grain sorghum. *Journal of Production Agriculture, 6*, 78–85.

Stewar, B. A., Musick, J. T., & Dusek, D. A., (1983). Yield and water use efficiency of grain sorghum in a limited irrigation-dryland farming system. *Agronomy Journal, 75*, 625–634.

Touchton, J. T., Gardner, W. A., Hargrove, W. L., & Duncan, R. R., (1981). Reseeding crimson clover as a N-source for no-tillage grain sorghum production. *Agronomy Journal, 74*, 84–96.

Wahua, T. A. T., & Mille, D. A., (1978). Effects of intercropping on soybean N_2-fixation and plant composition on associated sorghum and soybeans. *Agronomy Journal, 5*, 13–19.

Walter, Z. L., & Andrea, M., (2012). Are we ready to cultivate sweet sorghum as a bioenergy feedstock? A review on field management practices. *Biomass and Bioenergy, 40*, 1–12.

Wenceslau, G. T., Johannes, L., Winfried, E. H. B., & Wolfgang, Z., (2008). Nitrogen retention and plant uptake on a highly weathered central *Amazonian ferralsol* amended with compost and charcoal. *Journal of Plant Nutrition and Soil Science, 171*, 893–899.

CHAPTER 10

Harvest and Postharvest Technology

ABSTRACT

Grain quality of sorghum is affected by harvesting methods, post-harvest technology and other factors. Many research advancements have been made on harvest methods and the improvement of grain quality of sorghum that are presented in this chapter. A brief review of recent researches is presented to introduce the current state of the research on harvesting methods and on improving grain quality which can be reconsidered in new ways.

10.1 PHYSIOLOGICAL MATURITY

In sweet sorghum (*Sorghum bicolor* L. Moench) harvesting stage plays a pivotal role in determining the sugar content for the industrial production of alcohol. To ascertain the sorghum harvest growth stage for bioethanol production an experiment was conducted with four sweet sorghum genotypes in randomized block design (RCBD). Sorghum canes were harvested at intervals of seven days after anthesis (DAA). At different developmental stages (from flowering to physiological maturity) the genotypes were assessed for maximum bioethanol production. To produce ethanol the canes were crushed and the juice fermented. Further, chlorophyll content of harvested canes and weight of dried kernels at 14% moisture content were measured at different stages. Among genotypes and harvesting stages significant differences were observed for chlorophyll, grain weight, absolute ethanol volume, juice volume, cane yield, and brix. Findings unveiled that the sweet sorghum harvested at 104 to 117 days after planting (stages IV and V) are most suitable for the production of kernels and ethanol (Moses et al., 2017).

In northeastern Colorado, the shorter growing seasons because of low temperatures and early frost dates restricted the expansion of grain sorghum production from southeastern to northeastern Colorado. An experiment was performed to assess the possibility of grain sorghum attaining physiological maturity before the first frost fall. They used a simulation model to predict crop phenology (Phenology MMS). The past weather data was collected from nine sites for both irrigated and dry land phenological parameters. Based on this data physiological maturity for seven planting dates (1 May to 12 June), four seedbed moisture conditions influencing seedling emergence (from Optimum to Planted in Dust), and three maturity classes (early, medium, and late) were simulated. The plants grown under dryland conditions showed slightly higher chances of attaining maturity before the first frost. These changes reduced with an increase in latitude, longitude, and elevation, delayed planting date and late maturity classes. Therefore, the study findings provided producers with estimations of the dependability of growing grain sorghum in northeastern Colorado (Gregory et al., 2016).

In sorghum maturity gene loci that affect floral initiation was studied. Maturity gene loci were appeared to be a part of the floral regulatory system. Four gene loci regulate the time of floral initiation. Tropical varieties were dominant at locus 1 and generally at the other 3 loci also. Varieties from temperate zone were recessive at locus 1, or, if dominant at locus 1, they were recessive at either locus 2 or 4. Environment, particularly photoperiod and temperature had a greater influence on the expression of four genes. It was found that the earlier or late floral initiation is determined by the maturity genotype of the variety being cultivated. At the four gene loci the dominant and recessive alleles have significant interactions that have some bearing on the time of floral initiation. When locus 1 is dominant, dominance at the other three loci causes lateness. When locus 1 is recessive, dominance at locus 2 or both loci 2 and 3 causes earliness. At locus 1 heterozygosis has an unexpected interaction. If when locus 1 is heterozygous and locus 2 is recessive late floral initiation takes place. Delayed floral initiation was not observed when locus 2 is dominant and locus 1 is heterozygous. Heterozygosity at loci 2, 3, and 4 has no noticeable effect on the time of floral initiation (Quinby, 1967).

The effect of harvesting times on biomass yield, quality, and biomethane yield of sorghum was determined in a two-year field trial during 2016 and 2017. Harvesting times had a considerable effect on the growth characteristics, biomass quality and bio-methane yield. The lowest plant height, and dry matter (DM) yield were found at 60 DAS, whereas the highest values of these parameters were noted at 105 DAS. Likewise, the highest concentrations of protein and sugar were observed at 60 DAS, after that protein and sugar concentration were reduced significantly with progressing maturity. Furthermore, the maximum acid and neutral detergent fiber, lignin, cellulose, and hemicellulose were recorded at 120 DAS while; minimum acid detergent fiber, neutral detergent fiber, lignin, cellulose, and hemicellulose were found at 60 DAS. The highest specific methane yield was obtained at 60 DAS, whereas the lowest specific methane yield was found at 120 DAS. On the other hand, due to higher DM yield ha^{-1} maximum methane yield ha^{-1} was found at 105 DAS. Interestingly, a strong positive association was observed between DM yield and methane yield and a negative association between lignin concentrations and specific methane yield. The study determined that biomethane yield of sorghum can be increased by harvesting crop after 105 days from sowing as a result of high DM yield ha^{-1} (Figure 10.1) (Hassan et al., 2018).

10.2   CARES NEED TO BE TAKEN DURING HARVESTING

Sorghum is a vital crop of the developing world, and 60% of the crop being cultivated in Africa. Sorghum is harvested manually or by combine, before storage. Though storage structures and transport mechanisms are different, there is a possibility to further reduce post-harvest losses. The crop is mostly cultivated by smallholder farmers, mostly for domestic use. Commercial producers use trucks, rail, and barge for transportation of grain to domestic and foreign markets (Beta et al., 2016).

In arid and semi-arid lands, a sorghum grain has been adapted as an alternative staple crop. To determine post-harvest handling of sorghum grains a post-harvest handling farm survey was carried out. In six sorghum growing sub-counties (Siaya, Bondo, Njoro, Rongai, Kibwezi, and Kathonzweni), 88 farmers were sampled by means of a snowball sampling method. Data was collected on varieties, drying methods and storage forms. Using SPSS version 20 software descriptive statistics cross-tabulation the proportion of sorghum grains lost due to mold was examined. The results revealed that local varieties were more preferred by the farmers than the improved sorghum varieties. They stored sorghum grains in shelled form or on panicles. Sorghum mold incidence in the field and storage was main hazard to the sorghum grain in the sub-counties. Further, the research findings showed that farmers preserve a diversity of sorghum to cut down the percentage lost as storage duration of grains and biotic stress resistance. These findings can be used to elucidate unpredictable food insecurity in these sub-counties with prospective of sorghum production (Kange, 2016).

FIGURE 10.1   Sorghum crop at the harvesting stage.

The outer layer of the grain sorghum kernel contains wax. Present harvesting and handling methods bring about abrasion and breakage of the sorghum kernel, thereby reducing wax yield. In this study, the wax yield of sorghum was determined after mechanical harvesting and handling. The wax yield of the whole kernel was compared with that of broken sorghum kernels. Abrasion and breakage were caused by combine threshing and auger conveying of grain reducing the wax yield by 5%. In this study, a cleaning system was included as a monitored process that lessened the amount of broken kernels and sorghum particles, increasing wax yield after cleaning and bagging. Furthermore, the monitored process had a less amount of auguring and other handling involved (Lochte-Watson and Weller, 1999).

10.3 DRYING-MANAGEMENT OF GRAIN MOISTURE CONTENT

The researchers are looking for novel sources of cellulosic biomass to meet the world's energy needs. In this attempt the usage of new types of crops can bring about new processing problems. One problem that has been found is how to quickly process high-energy forage sorghum from a standing crop to a stable biomass package for transport and storage. Present commercial processing employs mechanical conditioners to increase the rate of water removal from plant cells. However, this type of equipment has been optimized for forage production instead of bioenergy feedstocks. This study examined three mechanical conditioner designs to find out the unit's power requirements and drying rate variations while processing high-energy forage sorghum. The three designs examined were fluted roll, chisel impeller, and "V" impeller. There was a non-significant difference in the power requirement of the three conditioners. The drying times for forage sorghum that was conditioned with the "V" impeller, chisel impeller and fluted roll conditioner was 43.2, 32.2, and 12.5 hours, respectively. Conditioning with chisel impeller, "V" impeller, and fluted roll reduced the drying time by 30.2, 47.8, and 79.7% than unconditioned material (Elizabeth et al., 2013).

An experiment was undertaken to find out the isothermal drying kinetic parameters of grain sorghum with a thermogravimetric analyzer (TGA). The kernels were placed in the TGA under isothermal drying conditions, i.e., 40, 50, 60, 70, 80, 90, and 100°C. Variations in the sample weight were ascertained from the TGA. The moisture ratio and the derivative weight loss curves were determined using the TGA data. Then the best-fit model for the experimental data was found by fitting the moisture ratio data to four well-known models, viz. Page, Newton, Logarithmic, and Henderson. The goodness of fit criteria was applied to find out the best-fit model. An increment in drying temperature from 40°C to 100°C speeded up the drying process and decreased the moisture ratio from 0.6091 to 0.2909, after 1 h. It was found that the Page model is best for 71.4% of the drying curves. However, Logarithmic, and Henderson models were the best fit for 28.6% of the studied cases. The effective moisture diffusivity increased from 0.96×10^{-8} m²/s to 1.73×10^{-8} m²/s with the increase in drying temperature from 40°C to 100°C. The drying activation energy value reached 9.4 kJ/mol under isothermal drying conditions (Sammy and Griffiths, 2018).

10.4   MARKETING THE HARVESTS

10.4.1   MARKETING OF SORGHUM IN INDIA

In India sorghum-marketing process is generally smooth and constraint-free. Sorghum can be easily marketed as food grain. Sometimes, linked or tied output and credit markets bring about distress sale by small and marginal farmers. The current marketing process may not be beneficial for the industries that use sorghum and millet. These industries may bypass the regular channels to acquire sorghum and millet with no trouble for a dual purpose: for food and industrial uses. The gradual commercialization and its integration of Indian agriculture with the world markets along with the increasing demand for sorghum, it's marketing, a former subsistence crop, assumes great significance. Though a substantial quantity of sorghum is traded, markets are still not well developed. Value-added and ready-to-use sorghum food products need not to be popularized. To promote the alternate uses of sorghum like in syrup, industrial products, jaggery from sweet sorghum, brewing of potable alcohol, and fuel alcohol from sweet stalks, new marketing channels need to be explored. The ordinary marketing model is not fit for the marketing of sorghum grain. Usually, only 8% to 9% of the crop carries over to the next marketing year. There are two main reasons for this low carry-over. First, due to the lower price of sorghum, it moves in and out of feeding channels as a substitute for maize in most markets. Second, the traditional market for sorghum is a replacement of the starch market. Starch is sold as binders for making wallboard and for production of ethanol. Ethanol production is one of the fastest-growing zones of the utilization of sorghum which now utilizes around 12% of the total sorghum grain production in the country (Dayakar, 2014).

10.5 RESEARCH ADVANCES IN POST-HARVEST TECHNOLOGY

10.5.1 SEED CHEMICAL COMPOSITION

Linda et al. (2006) analyzed the chemical composition of a variety of sorghum types. They found that sorghum grains had varying levels of phenolic acids, flavonoids, and condensed tannins. Similar to fruits and vegetables, in some of the pigmented varieties of sorghum the antioxidant levels were found to be very high. Sorghum grain though a good source of phenolic compounds, it was found that not all the sorghums contained condensed tannins, though phenolic acids were found to be high in all the sorghum grains. In addition, unique anthocyanins were detected in some of the pigmented sorghums. These pigmented sorghums because of their unique anthocyanin contents could have the potentiality of being used as food colorants. A few sorghums were found to possess a major pigmented testa. This pigmented testa contained condensed tannins, mostly of flavan-3-ols with variable length. Further, they separated and quantitatively analyzed the Flavan-3-ols of up to 8–10 units. Sorghums with these tannin contents can act as good antioxidants. High concentration of 3-deoxyan-thocyanins (i.e., luteolinidin, and apigenidin) in sorghums gives stable pigments at high pH. Pigmented and tannin sorghum varieties also had high antioxidant levels as that of fruits and vegetables.

Christian et al. (1991) compared the lignin structure and enzyme activities of normal and brown-midrib (BMR-6) mutant lines of *Sorghum bicolor* and identified the enzyme(s) participating in the reduction of the lignin content of the mutant. They found that there is a depression in the activity of cinnamyl-alcohol dehydrogenase (CAD) and caffeic acid O-methyltransferase (COMT) in BMR-6 line. However, it was also noted that this reduction in the activity of CAD corresponded mainly to the structural modifications. Thus, it was observed that this apparent depression in the activity of cinnamyl alcohol dehydrogenase (CAD) activity resulted in a variation in the *Sorghum* lignin content, affected by depression of CAD activity. This decline in the CAD activity is additionally accompanied by the incorporation of other chemicals like cinnamaldehydes into core lignin.

The analysis of antioxidant property of sorghum brans, baked, and extruded products by oxygen radical absorbance capacity (ORAC), 2,2'-azinobis (3-ethyl-benzothiazoline-6-sulfonic acid) (ABTS), and 2,2-diphenyl-1-picrylhydrazyl (DPPH) have shown that ABTS and DPPH methods were more economical and simpler predictive power (Joseph et al., 2003).

Sorghum is a foremost raw material for starch and ethanol industry. The use of improved hybrids/varieties for specific purpose would make sorghum more remunerative to farmers and industries. The identification of germplasm and genetics of the character is essential for any successful breeding program. Here, the variability for starch (quantity and quality) and ethanol yield of sorghum were deliberated. Further, the traits related to the inhibition of ethanol production and high ethanol recovery was reported. Dried distiller's grain with solubles (DDGS) is one of the coproducts in the starch/ethanol industry. Sorghum lines vary in DDGS quality particularly for crude protein, lysine content, and crude fat content. Sorghum-DDGS contains a valuable health-promoting compound, policosanols, which is absent in maize. Increasing grain size of sorghum hybrids would help in wet-milling procedure as germ can be easily removed from bold grain (Audilakshmi and Swarnalatha, 2019).

Sweet sorghum syrups are renewable raw material available throughout the year for the production of biofuels and biochemicals. Though sweet sorghum sugars have been used for the production of butanol in the past, most of the researches concentrated on sweet sorghum juice and not on sweet sorghum syrups. Therefore, in this study the viability of using the syrups as feedstock was investigated. Previous reports revealed that sweet sorghum syrups diluted to 60 g/L of glucose equivalents, could not be used as a direct substitute for a synthetic growth medium for industrial butanol-producing strain *Clostridium beijerinckii*. Additional studies showed that supplementary nutrients particularly, ammonium-nitrogen is needed for effective fermentation. This was applicable for both production sources of sweet sorghum syrups from commercial cultivars and hybrids. Usually, these two sources produced 15 g/L of total acetone, butanol, and ethanol with about half of that being butanol. Insignificant statistical differences were observed between the production potential of the two sources. Never-theless, aconitic acid, which was found in the same levels in both syrups, was ruled out as a butanol fermentation inhibitor at the fermentation pH > 4.5 (Klasson et al., 2018).

Biological pretreatment of lignocellulosic biomass is thought to be energy-efficient and cost-effective. In this study, a white-rot fungus *Phanerochaete chrysosporium* (MTCC 4955) was used for the biological treatment of sorghum husk under submerged static conditions. Ligninolytic enzymes like lignin peroxidase (0.843 U/mL) and manganese peroxidase (0.389 U/mL) performed an important role in the biological pretreat-ment of sorghum husk. During the biological pretreatment of sorghum husk activities of various hydrolytic enzymes such as endoglucanase (57.25 U/mL), exoglucanase (4.76 U/mL), filter paperase (0.580 U/mL), glucoamylase (153.38 U/mL), and xylanase (88.14 U/mL) were evaluated. The reducing sugar produced from enzymatic hydrolysis of untreated sorghum husk was 20.07 mg/g, whereas reducing sugar produced from biologically pretreated sorghum husk was 103.0 mg/g. This indicated a significant increase in the production of reducing sugar in the biologically pretreated sorghum husk when compared to its untreated counterpart. Further, biologically pretreated sorghum husk hydrolysate was fermented for 48 h using *Saccharomyces cerevisiae* (KCTC 7296), *Pachysolen tannophilus* (MTCC 1077), and their co-culture; this resulted in ethanol yields of 2.113, 1.095, and 2.348%, respectively. After the delignification and hydrolysis the surface characteristics of the substrate were assessed with the help of FTIR, XRD, and SEM, which confirmed the effectiveness of the biological pretreatment process (Waghmare et al., 2018).

The interest in stable natural colorants for food uses continues to grow. In West Africa, a red pigment extracted from the leaf sheaths of a sorghum variety having high content of apigeninidin is extensively utilized as a biocolorant in processed foods. Here, the potential for improvement and utilization of the red biocolorant from dye sorghum in the food sector was determined by comparing the color and anthocyanin composition from conventional extraction methods. Sorghum bio colorant was commonly used in fermented and heated foods. Conventional extraction methods differ in two aspects, viz. the usage of an alkaline rock salt and the temperature of the extraction water. In the extraction of anthocyanins cool extraction using the alkaline ingredient was found to be more effective than hot alkaline and hot aqueous extractions. The cool and hot alkaline extracts had three times higher apigeninidin content than the aqueous extract. Therefore, the study concluded that the apigeninidin can be efficiently extracted from dye sorghum leaf sheaths using cool and hot alkaline extraction methods at pH 8–9. Wider use of the sorghum biocolorant in foods needs additional studies on its effects on bioavailability of nutrient and antioxidant activity (Akogou et al., 2018).

The use of bioethanol as biofuel may lead to reductions in greenhouse gases and gasoline imports. Further, it can replace the air and underground water pollutants such as lead or MTBE (Methyl tert-butyl ether) respectively. Plants are the best source for the extraction of bioethanol. For this purpose, a comparative analysis of the technological options with diverse feedstocks should be made. Previous studies and this research indicated that sweet sorghum can be used as a feedstock for the production of ethanol under hot and dry climatic conditions. Owing to its higher tolerance to salt and drought when compared to sugarcane and corn that are presently used for biofuel production in the world. Additionally, the sweet sorghum requires much lower water and fertilizer than sugarcane and the carbohydrates content of sweet sorghum stalk is similar to sugarcane. In addition, the high fermentable sugar content in sweet sorghum stalk makes it to be more appropriate for ethanol fermentation. Therefore, it was proposed that the use of sweet sorghum for biofuel production in hot and dry countries could solve many problems such as increasing greenhouse gases and gasoline imports (Almodares and Hadi, 2009).

Lipid-soluble secondary plant metabolites like carotenoids and tocochromanols are crucial for the normal functioning of plants and, sometimes they serve as a source of vitamin A and E in humans. In developing countries with increasing interest in biofortification of grains to decrease micronutrient deficiencies improvement of the provitamin A carotenoid and tocochromanols levels in sorghum and other cereal grains by means of conventional breeding and transgenic methods has increased. With this advancement in research, consistent methodology for the extraction, identification, and quantification of individual carotenoids and tocochromanols species from sorghum and other cereal grains is vital. In this paper, Ortiz et al. (2019) described a simple method for extraction of carotenoid and tocochromanols adapted to sorghum grain and chromatographic conditions for separation, identification, and quantification of individual carotenoid with high-performance liquid chromatography (HPLC) system.

Sorghum starches properties have been modified by using heat-moisture treatment (HMT). However, much information is not available on the effects of HMT on the starch digestibility and the simultaneous variations in sorghum flour protein. In this study, heat-moisture conditions to improve the resistant starch (RS) content of sorghum flour were identified and variations in sorghum proteins and starch structure were studied. Sorghum flours with different moisture contents (0, 125, 200, and 300 g kg^{-1}w.b.) were heated at three temperatures (100, 120, and 140°C) and times (1, 2, and 4 h). RS level was increased by HMT of sorghum flour. A high RS content (221 g kg^{-1} vs. 56 g kg^{-1} for the untreated flour) was recorded in the flour treated at 200 g kg^{-1} moisture and 100°C for 4 h. When sorghum flour was heated at moisture content of 200 g kg^{-1} or below starch was not gelatinized. During HMT reduction in sorghum protein digestibility and solubility were recorded. The increase in RS of sorghum flour upon HMT was accredited to improved amylose-lipid complexes and heat-induced structural variations in its protein fraction. The study concluded that HMT can be used to improve RS content in sorghum flour without gelatinization of its starch, thus offering sorghum flour with distinctive food applications (Vu et al., 2017).

A study was carried out to examine the effect of malting and fermentation on the ant nutritional component and functional characteristics of sorghum flour. Cleaned sorghum grain was milled to pass through 40-mesh sieve to get whole sorghum flour. Then cleaned sorghum grain was steeped in 0.2% calcium hydroxide solution for 24 hr for malting and germinated for 48 hr at 90% RH and 27 ± 2°C. Sprout was taken out, dried in a hot air oven at 50 ± 2°C for 24 hr and milled to pass through 40 mesh sieves. To get fermented sorghum flour, cooked (88 ± 2°C) sorghum flour was mixed with 13.3 mg% diastase and 2 mg% pepsin (on the basis of whole sorghum flour weight) and left for 1 hr. *Lactobacillus plantarum* (10 7 cfu/g) was inoculated and incubated at temperature 30 ± 2°C for 48 hr. The fermented slurry was dried at 50 ± 2°C in a hot air oven for 24 hr and milled to pass through 40-mesh sieve. The yield of sorghum flour was lower than whole and malted sorghum flour. Sorghum germination reduced phytate, tannin, and oxalate by 40%, 16.12% and 49.1%, respectively, whereas fermentation of sorghum flour reduced above by 77%, 96.7% and 67.85%, respectively. The hydrogen cyanide in malted sorghum flour and whole sorghum flour was the same, but fermentation of sorghum flour reduced hydrogen cyanide by 52.3%.

The malting and fermentation reduced the bulk density and viscosity significantly reduced, but they increased the water absorption capacity and oil absorption capacity. Despite lower yield, fermented flour was good because of reduced ANF and improved functional property (Ojha et al., 2018).

An experiment was undertaken to assess the variations in sweet sorghum (*Sorghum bicolor* L. Moench) after harvest. Three sorghum cultivars were harvested at 90, 115, and 140 days after planting. Stripped and topped stalks were separated into four treatments: whole stalks, 20- or 40-cm billets, and chopped. Samples were kept in individual plastic bins and stored outside in a shade tent. At 0, 1, 2, and 4 days after harvest, subsamples were taken out from each bin and the juice expressed. °Brix (percent total dissolved solids on a w/w basis) and the simple sugars glucose, fructose, and sucrose were analyzed in juice. During storage slight reduction or no change in juice °Brix was observed in most of the treatments. In all treatments reduction in sucrose and increment in glucose and fructose in the whole stalk, 20-cm billet, and 40-cm billet treatments was recorded. Over the 4-day storage period total sugar changed slightly in those treatments. There were significant decreases in glucose in chopped sorghum after 1 day of storage. Glucose reduced to near 0 mg mL^{-1} by 2 days after harvest, while the reduction in fructose was not significant. The correlation between total sugar and Brix using day 0 means (non-deteriorated juice) was r$=0.964$ (n$=32$, $\rho<0.001$). The correlation between total sugar and Brix using the shredded sample means, which included deteriorated juice, was r$=0.411$ (n$=24$, $\rho \leq 0.046$). The study established as forage-harvested sorghum would have to be processed within hours to preserve the sugars, whole stalk and billet harvesting were superior to harvesting by forage harvester (Sarah et al., 2012).

Whole kernels of sorghum were separated into bran and abraded kernel fractions through abrasive decortication of grain sorghum concentrate the source of wax. A tangential abrasive dehulling device (TADD) was used for abrasive decortication. The variables studied were moisture addition with tempering time and abrasion time. The components of DM, starch, protein, ash, and wax were assessed for whole sorghum kernels. After decortication mass of abraded kernels and branfractions were also measured. Then chemical analysis was made to ascertain the amount of each component recovered in each fraction. A summation of the amounts recovered in each fraction was used to determine the total matter recovered. In this study, the

TADD method recovered an average of over 94% of the total DM, total starch, total protein and total ash between the two fractions, whereas less than 85% of the total wax was recovered from any one sample. As longer abrasion times were used to remove the bran from the kernel, more starch was removed as well. After abrading kernels at 10.4% moisture content for 80s maximum wax concentration, $66.9 \pm 5.6\%$, in the bran fraction with a starch content of $5.5 \pm 0.8\%$ occurred. Finally, minimal starch contamination was not achieved using the TADD to concentrate wax to high levels (Lochte-Watson et al., 2000).

The effect of post-harvest handling (threshing and storage) methods on the quality of sorghum (*Sorghum bicolor* (L) Moench) grains variety Numbu was investigated. The parameters considered were percentages of damaged grains and seed germination, population growth of *S. zeamais*, *F. proliferatum,* and *F. verticillioides*; fumonisin B_1 and carbohydrate contents, and the percentage of weight loss during storage. The variation in moisture contents of sorghum grains was also recorded. A stick of wood and a paddy thresher was used for threshing. Sorghum grains were packed in hermetic plastic bags. Each bag with different conditions inside of the bag was introduced with 10 pairs of *Sitophilus zeamais* (1–14 days old). Under warehouse conditions sorghum was stored for one, two, and three months. The results revealed, that sorghum had lower moisture content during storage than that its standard safe moisture content (<14%). At the beginning of storage, the percentage of damaged grains caused by threshing using a stick of wood was higher than that of using a paddy thresher. During storage the increase of *S. zeamais* population and the grains stored under normal conditions resulted in the increased percentage of damaged grains among others accordingly, the percentage of weight loss was also increased. Sorghum threshed using a stick of wood showed lower seed germination percentage and higher fumonisin B_1 content than that of threshed using a paddy thresher. The percentage of seed germination and the population of *F. proliferatum* and *F. verticillioides* decreased with the increasing storage duration. Generally, there was no significant difference in the carbohydrate content of sorghum threshed with a stick of wood and a paddy thresher from at the initiation of storage up to three months of storage. Threshing using a paddy thresher was found to be better than threshing using a wooden stick (Okky et al., 2012).

10.6   CONCLUSIONS

The yield and grain quality of sorghum can be enhanced by harvesting it at the appropriate time and by taking care during harvesting. In this chapter, an overview of the suitable harvesting time and care to be taken during harvesting was discussed. The post-harvest uses of sorghum for the production of bioethanol and other products are also discussed.

KEYWORDS

- cinnamyl-alcohol dehydrogenase
- dry matter
- harvesting
- post-harvesting technology
- sorghum
- thermogravimetric analyzer

REFERENCES

Akogou, F. U. G., Kayodé, A. P., Polycarpe, D. B., Heidy, M. W., & Linnemann, A. R., (2018). Extraction methods and food uses of a natural red colorant from dye sorghum. *Journal of the Science of Food and Agriculture, 45,* 76–84.

Almodares, A., & Hadi, M. R., (2009). Production of bioethanol from sweet sorghum: A review. *African Journal of Agricultural Research, 4,* 56–67.

Audilakshmi, S., & Swarnalatha, M., (2019). Sorghum for starch and grain ethanol. In: C. Aruna, K.B.R.S. Visarada, B. Venkatesh Bhat, Vilas A. Tonapi (eds). Woodhead Publishing Series in Food Science, Technology and Nutrition, Breeding Sorghum for Diverse End Uses. Woodhead Publishing, pp. 239-254. ISBN 9780081018798.

Beta, T., Chisi, M., & Monyo, E. S., (2016). Sorghum: Harvest, storage, and transport. *Encyclopedia of Food Grains, 4,* 54–61.

Christian, P., Marcel, M. M., Jaap, J. B., Birgit, F., & Andres, B., (1991). Involvement of cinnamyl-alcohol dehydrogenase in the control of lignin formation in *Sorghum bicolor* L. Moench. *Planta, 185,* 538–544.

Dayakar, R. B., (2014). Marketing of sorghum. *Foundation Books,* 19–26.

Elizabeth, A. M., Michael, D. B., Raymond, L. H., & Randal, K. T., (2013). *Machine Power and Drying Rate Relationships for High Energy Forage Sorghum* (Vol. 7, pp. 56–63). Published by the American Society of Agricultural and Biological Engineers, St. Joseph, Michigan, Kansas City, Missouri.

Gregory, S. M. M., Debora, A. E., Sally, M. J., Jerry, J. J., & Merle, F. V., (2016). Simulating the probability of grain sorghum maturity before the first frost in Northeastern Colorado. *Agronomy, 31,* 11–18.

Hassan, M. U., Chattha, M. U., Chattha, M. B., Mahmood, A., & Sahi, S. T., (2018). Impact of harvesting times on chemical composition and methane productivity of sorghum (*Sorghum bicolor* moench L.). *Applied Ecology and Environmental Research, 16,* 2267–2276.

Joseph, M. A., Lloyd, W. R., Xianli, W., Ronald, L. P., & Luis, C. Z., (2003). Screening methods to measure antioxidant activity of Sorghum (*Sorghum bicolor*) and sorghum products. *Journal of Agriculture and Food Chemistry, 51,* 6657–6662.

Kange, A. M., (2016). Assessment of handling sorghum (*Sorghum bicolor* L. Moench) grains in production areas. *International Journal of Plant and Soil Science, 10,* 1–7.

Klasson, K., Thomas, K., Qureshi, Nasib, P., Randall, H., Matthew, E., & Gillian, (2018). Fermentation of sweet sorghum syrup to butanol in the presence of natural nutrients and inhibitors. *Sugar Technology, 5,* 78–86.

Linda, D. L., & Rooney, W., (2006). Sorghum and millet phenols and antioxidants. *Journal of Cereal Science, 44,* 236–251.

Lochte-Watson, K. R., & Weller, C. L., (1999). Technical notes: Wax yield of grain sorghum (*Sorghum bicolor*) as affected by mechanical harvesting, threshing, and handling methods. *Applied Engineering in Agriculture, 15,* 69–72.

Lochte-Watson, K. R., Curtis, L. W., & David, S. J., (2000). PH-Postharvest technology: Fractionation of grain sorghum using abrasive decortication. *Journal of Agricultural Engineering Research, 77,* 203–208.

Moses, O. O., James, O. O., Maurice, E. O., Erick, C., Betty, M., & Justice, R., (2017). Effect of harvesting stage on sweet sorghum (*Sorghum bicolor* L.) genotypes in Western Kenya. *Scientific World Journal, 15,* 1–10.

Ojha, Pravin, A., Roshan, K., Roman, M., Achyut, S., Ujjwol, K., & Tika, B., (2018). Malting and fermentation effects on antinutritional components and functional characteristics of sorghum flour. *Food Science and Nutrition, 5,* 34–41.

Okky, S. D., (2012). Post-harvest quality improvement of sorghum (*Sorghum bicolor* (L.) Moench) *grains. The Southeast Asian Journal of Tropical Biology, 19,* 255–263.

Ortiz, D., Darwin, F., & Mario, G., (2019). Identification and quantification of carotenoids and tocochromanols in sorghum grain by high-performance liquid chromatography. *Methods in Molecular Biology, 5,* 123–134.

Quinby, J. R., (1967). The maturity genes of sorghum. *Advances in Agronomy, 19,* 267–305.

Sammy, S., & Griffiths, A., (2018). Grain sorghum drying kinetics under isothermal conditions using thermo gravimetric analyzer. *Bio Resources, 13,* 1–20.

Sarah, E. L., Thomas, L. T., Hrvoje, R., & Deborah, L. B., (2012). Post-harvest changes in sweet sorghum I: Brix and Sugars. *Bioenergy Research, 5,* 158–167.

Vu, T., Thanh, H. B., Scott, H., Chao, F. S., & Yong, C., (2017). Changes in protein and starch digestibility in sorghum flour during heat-moisture treatments. *Journal of the Science of Food and Agriculture, 3,* 45–52.

Waghmare, P., Pankajkumar, R., Khandare, R. V., Jeon, Byong, H. G., & Sanjay, P., (2018). Enzymatic hydrolysis of biologically pretreated sorghum husk for bioethanol production. *Biofuel Research Journal, 3,* 56–67.

CHAPTER 11

Sorghum Grain Quality Analysis, Food Quality Characteristics, Chemistry, and Food Processing

ABSTRACT

Several research works were carried out to improve sorghum grain quality through various methods, and many advancements have been achieved that need to be assembled at one place. Therefore, in this chapter, recent approaches and their effectiveness in determination of grain characteristics and food quality are presented.

11.1 SORGHUM GRAIN QUALITY ANALYSIS

Sorghum is of most important cereal crop in the world and is a source of nutrients and bioactive compounds for the human diet. The current findings regarding the nutrients and bioactive compounds of sorghum and their probable influence on human health were summarized. They analyzed the drawbacks and positive points of the reports to propose directions for forthcoming research. Sorghum mostly contains starch, which has low digestibility proteins and unsaturated lipids, therefore less digestible than that of other cereals. It is a source of various minerals and vitamins. Moreover, most sorghum varieties are rich in phenolic compounds, particularly 3-deoxyanthocyanidins and tannins.

Sorghum (*S. bicolor* L. Moench) is a staple food in several parts of the world and ranks fifth as a cereal crop in terms of production and consumption. However, the food quality of sorghum has not yet been defined clearly, maybe due to its limited use in commercial foods unlike wheat, rice, and maize. In areas where sorghum is consumed, only small quantities of products appear in metropolitan markets. There are few standards available to distinguish grain quality, which is assessed mainly by subjective criteria such as kernel color, appearance, size, and shape (Rooney et al., 1981).

Rat balance tests were undertaken to determine the nutritional quality of whole and decorticated sorghum grains (Tetron, Dabar, Feterita) collected from Sudan. Different dishes were prepared using grains from these varieties and analyzed. It was found that the true digestibility of the protein got reduced when porridge (Ugali) was cooked from the Dabar (low polyphenol) variety; however, biological value increased when compared with the raw grain. These variations were not seen were when pH was adjusted to 3.9 before cooking. More pronounced effects of cooking were observed in the variety of Feterita (high polyphenol). When pancakes (Kisra) were fermented at pH 3.9, negligible effects of cooking on nutritional quality were observed in all the sorghum varieties studied. Somewhat less digestible energies and lower true protein digestibilities were recorded in cooked fermented and cooked unfermented, acid-adjusted Aceda (a thin fermented gruel). However, these dishes had higher biological values, than uncooked fermented Aceda, indicating that the protective effect of acidification noted with Ugali was not due to Aceda. A higher biological value was observed in Marissa, a sorghum beer, after sieving than did the unsieved brew. However, Marissa preparation had a negative effect on true protein digestibility. The test results unveiled that sorghum is low in lysine, and thus, has a low biological value (Eggum et al., 1983).

In the wheat industry, the single kernel characterization system (SKCS) has been extensively used, and its parameters were related to end-use quality in wheat. However, SKCS was proposed to examine wheat, which has a different kernel structure from sorghum; it can be used to evaluate sorghum grain quality. To understand the meaning of SKCS estimates for grain sorghum length, width, thickness (diameter) and weight of individual sorghum grains were measured using laboratory methods and SKCS. SKCS estimates for kernel weight and thickness were highly correlated to laboratory measurements. However, for kernel thickness SKCS estimates were undervalued by ≈20%. Seven sorghum samples with variable hardness values were tempered to four moisture levels to evaluate the SKCS moisture prediction for sorghum. When moisture contents estimated by SKCS were compared with a standard oven method, the SKCS moisture estimates were found to be less than moisture measured by air oven, particularly at low moisture content. Further, SKCS hardness values and hardness measured by abrasive decortication were compared and a moderate correlation was noted between them. The study concluded that the SKCS is appropriate for measuring sorghum grain characteristics. However, additional study is required to ascertain how SKCS hardness estimates are associated with milling properties of sorghum grain (Bean et al., 2006).

Grain molds caused by a complex of fungi is a major cause of sorghum grain quality loss during rainy season. It was found that some traits such as glume color, glume cover, and grain hardness may provide resistance against grain mold in white grain sorghum. Therefore, a study was conducted to work out the nature and degree of gene effects for grain hardness. Sorghum line with soft grain, 278 and a hard and grain mold-resistant line, 858,586 were crossed and evaluated by generation mean analysis in two seasons. The grinding time needed to get a fixed volume of flour from the grains was measured to record the hardness of sorghum grain. The distribution of F_2 frequency exhibited continuous variation, which indicated polygenic nature of grain hardness inheritance. The generation mean analysis revealed the preponderance of additive gene effects and additive × additive gene interactions. Therefore, the direct selection can be used to increase frequency of desirable genes in sorghum (Aruna and Audilakshmi, 2004).

Heavy rains during sorghum crop maturity causes serious deterioration of grain quality. The main reasons for this deterioration are grain mold caused by complex of fungi and leaching out of color from glume, i.e., discoloration of the grain. It was found that this poor quality of the grain is the main reason for the stagnation of the area and production of sorghum during rainy season in India. In this chapter, different methods to control the mold problem were explored in 18 farmers' fields each in the districts of Parbhani, Akola (Maharashtra), Mahabubnagar (Andhra Pradesh), Indore (Madhya Pradesh), Coimbatore (Tamil Nadu) and Dharwad (Karnataka) of India: (i) varieties with superior grain quality were identified among the released varieties; (ii) influence of anti-heating chemicals and fungicides on grain quality in relation to molds; (iii) the improvement of deteriorated grain by pearling; (iv) identification of varieties tolerant to grain mold; and (v) solarization to increase storability of rainy season grain. Among the released varieties, good quality grains with tolerance to mold were found in CSH 16. It's bold, round, and lustrous grain raised its market price by 21% than those of other cultivars. Among the anti-heating chemicals, acetic acid was more effective against mold. Pearling of grain also increased its market price. Under epiphytotic conditions the high yielding variety SVD 9601 exhibited tolerance to grain mold during all the three years. The solarization and storing of harvested produce in metal bins reduced the infestation of insect by about 40%. These technologies can be followed either separately or in combination. Thus, the best package of technology to improve the grain quality of the rainy season produce would be cultivation of good quality high yielding varieties followed by the harvesting of produce at physiological maturity, artificial drying, and storing the solarized produce in metal bins (Audilakshmi et al., 2007).

Sorghum (*Sorghum bicolor* L. Moench) is a main source of energy and protein for millions of the world's poorest people. The low digestibility of grain protein and starch diminished its nutritional value. To deal with this problem, the properties of two sorghum lines (KS48 and KS51) with common pedigree but different digestibility were analyzed. The proteins and starch of both the lines were digested using *in vitro* assays on the basis of pepsin and α-amylase. The protein and starch of KS48 was digested more thoroughly than KS51. These results were consistent with the results of ruminal fluid assay. It was found that the large quantity of disulfide-bonded proteins, presence of non-waxy starch and the associated granule-bound starch synthase and the different nature of the protein matrix and its interaction with starch were main reasons for the indigestibility of KS51. Thus, the study findings suggested that each of these factors should be taken into account in attempts to improve the nutritional value of sorghum grain (Joshua et al., 2009).

11.2   GRAIN CHEMISTRY

Sorghum and other millets (pearl, proso, foxtail, finger, barnyard, kodo, fonio, and teff) are most important source of calories and protein for millions of people in Africa and Asia. Sorghum is also a main fodder grain in other countries. These grains have different chemical-nutritional and structural properties, which are influenced by both genetics and environmental conditions. Further, they differ in appearance, grain physical properties, starch composition, essential amino acid composition or protein quality, fatty acid composition, and phytochemical profiles at intraspecies level. Particularly, sorghum has been categorized according to color and presence or absence of condensed tannins. Three major anatomical parts of sorghum and millets grains are pericarp (bran), germ, and endosperm, which vary in their chemical composition (Sergio et al., 2019).

To determine the amylase activity, nutritional composition/properties and ability of sorghum to meet the Recommended Dietary Allowance weaning food was made from sorghum and cowpea based on a malted technology. Through steeping, germination, drying, toasting, grinding, and sieving malted sorghum flour and steamed cooked cowpea flour was made. Both flours were mixed in 2:1 ratio and malted-weaning food (GSC), unmalted sorghum and steamed cooked cowpea was made (USC). The optimum amylase activity of sorghum was ascertained. Owing to reduction in viscosity, optimum amylase activity with reduced dietary bulk in GSC was recorded after 72 hours. Germination did not show any significant effect on protein contents, 12.07 g (GSC) and 12.57 g (USC), samples met 1/3 RDA protein requirement for 1–3 years old child. However, there was significant increment in essential amino acids excluding sulfur amino acids and tryptophan. Amino acid value for GSC was approximately equal to FAO reference pattern except for threonine that was also the limiting amino acid. In GSC an increment in vitamin A, i.e., beta-carotene (267.0 IU/100 g), thiamin (0.24 mg/100 g) and ascorbic acid (5.0 mg/100 g) from 197.0 IU/100 g (Vit. A), 0.16 mg/100 g (thiamin) and 2.73 mg/100 g (ascorbic acid) was observed. Phosphorus and iron contents also increased from 91.65 mg/100 g and 4.01 mg/100 g to 100.0 mg/100 g and 6.40 mg/100 g, respectively. Nutritional values of weaning food prepared from germinated sorghum were higher than ungerminated. Germination significantly increased essential amino except for histidine, sulfur amino acid and tryptophan. It also increased phosphorus, iron, vitamin A (β-carotene) thiamine, riboflavin, niacin, and ascorbic acid (Elemo et al., 2011).

The results acquired *in vitro* and in animals revealed that phenolics compounds and fat soluble compounds (polycosanols) isolated from sorghum benefit the gut microbiota and parameters associated with obesity, oxidative stress, inflammation, diabetes, dyslipidemia, cancer, and hypertension. The effects of whole sorghum and its fractions on human health need to be assessed. To put it briefly, sorghum is a source of nutrients and bioactive compounds, particularly 3-deoxyanthocyanidins, tannins, and polycosanols, which beneficially modulate, *in vitro* and in animals, parameters associated to noncommunicable diseases. Research should be directed to assess the effects of different processing on protein and starch digestibility of sorghum as well as on the profile and bioavailability of its bioactive compounds, particularly 3-deoxyanthocyanidins and tannins. Additionally, the advantages resultant from the interaction of bioactive compounds in sorghum and human microbiota should be examined (De Morais et al., 2017).

Seeds from over 9,000 sorghum lines were evaluated for probable increases in lysine concentration. Sorghum lines were classified based on endosperm phenotype to select floury endosperm lines. Sixty-two selected floury endosperm lines were examined for protein and lysine composition. Higher lysine content was observed in two floury lines of Ethiopian origin, IS 11167 and IS 11758 at relatively high protein levels. At 15.7 and 17.2% protein, the average whole-grain lysine content of these lines was 3.34 and 3.13 (g/100 g protein), respectively. Additionally these lines had high oil percentage. Carbohydrate content in whole-grain samples of two high lysine lines was comparable to that of normal sorghum grain. However, high lysine lines showed a double increase in sucrose concentration. In comparison to normal endosperm checks the high lysine gene altered the amino acid pattern in **hl hl hl** endosperm tissue. Increase in lysine, arginine, aspartic acid, glycine, and tryptophan concentrations and reduced quantities of glutamic acid, proline, alanine, and leucine were major changes observed in **hl hl hl** endosperm. Inheritance studies unveiled that a single recessive gene regulates high lysine concentration of each line, though it is not identified whether the genes from both lines are allelic. The high lysine genes located in Ethiopian lines IS 11167 and IS 11758 were termed as hl. The hl gene homozygous lines showed partially dented endosperm of kernels. The biological value of high lysine lines was much higher than normal sorghum lines. A 28 days isonitrogenous feeding experiment was performed. The results showed that the weight gain of weanling rats was three times higher on an IS 11758 ration and twice on an IS 11167 ration when compared with the weight gains on rations made from normal sorghum lines. When rats were fed with rations without any dilution except the usual 2% vitamin and 4% mineral supplementation, the weight gain was 94 g on high lysine sorghum (IS 11758) and 28.5 g on best nutritional quality sorghum line (IS 2319), against 91.5 g on opaque-2 corn (*Zea mays* L.) and 30.2 g on normal corn. In this 28 days trail feed efficiency ratios were 3.0 for high lysine sorghum, 6.8 for IS 2319, 3.4 for opaque-2 corn, and 7.4 for normal corn (Rameshwar and John, 1973).

A study was carried out to analyze the effect of chemical treatments on polyphenols and malt quality in sorghum. Two sorghum lines Chirimaugute and DC-75 with tannin and one tannin-free sorghum, SV2, were soaked in water, HCl (0.25 m), formaldehyde (0.017 m) and NaOH (0.075 m) for 8 and 24 h. Germination was carried out for 2 and 5 d. The grains soaked in NaOH showed greater water uptake than other treatments. The polyphenol content of the raw grain get reduced in all treatments. Treatment with NaOH or formaldehyde (HCHO) was found to be more effectual than water or HCl. Malt quality was measured as diastatic power (DP). Potential DP ascertained after the extraction of peptone, revealed higher amylase activity in DC-75 malt than in Chirimaugute and SV2 malt. Available DP, ascertained after water extraction, was low in malt from the tannin-containing varieties treated with water or HCl. Only malting cannot decrease the enzyme inhibitory power of the sorghum tannins. The NaOH and HCHO treatments of the tannin varieties markedly improved the available DP. The study concluded that soaking of sorghum grains in dilute NaOH is helpful in detoxifying high-tannin sorghums. The reduction in steeping period enhances malt quality. Soaking in NaOH seems to be a safer substitute to HCHO to treat high-tannin sorghums in the malting industry and for other food applications (Betaa et al., 2000).

Duodu et al. (2002) studied the influence of grain structure and cooking on *in vitro* protein digestibility (IVPD) in sorghum and maize. The improvement in *in-vitro* protein digestibility of uncooked and cooked sorghum was observed with the reduction in structural complexity of the sample from whole-grain flour through endosperm flour to protein body enriched samples. However, this was not observed for maize. Cooking decreased protein digestibility of sorghum but not maize. When cooked maize and sorghum whole-grain and endosperm flours were treated with alpha-amylase to lessen sample complexity prior to *in vitro* pepsin digestion protein digestibility was improved slightly. It was found that the total polyphenol content of samples was not associated with the decline in sorghum protein digestibility on cooking. They identified pericarp components, germ, endosperm cell walls, and gelatinized starch as probable factors affecting digestibility of sorghum protein. Under non-reducing conditions electrophoresis of uncooked and cooked protein body-enriched samples of sorghum and maize, and prolamin fractions of sorghum revealed oligomeric proteins with molecular weights (Mr) 45, 66 and >66 kDa and monomeric kafirins and zeins. When compared to maize more

45–50 kDa oligomers were found in protein-body-enriched samples of sorghum. In cooked sorghum, some of these oligomers showed resistance to reduction. Pepsin-indigestible residues from protein-body-enriched samples had α-zein (uncooked and cooked maize) or α-kafirin (uncooked sorghum). In Addition to these β- and γ-kafirin and reduction-resistant 45–50 kDa oligomers were found in cooked sorghum. It was appeared that cooking forms disulfide-bonded oligomeric proteins that occur to a greater amount in sorghum than in maize. This may elucidate the poorer protein digestibility of cooked sorghum.

The endogenous polyphenols found in the testa and nucellar layer of the sorghum grain did not show any effect on the enzyme production and activity in malt produced from bird proof sorghum cultivars. When milling of sorghum grain disrupted the rigid segregation of tissues and substances in the malt grain, the polyphenols inhibited the endogenous enzymes in aqueous suspensions and reduced the brewing value of the ground malt. Sorghum beer prepared from malt of bird proof cultivars was not bitter in taste. Bird proof sorghum cultivars exhibited wide differences in amount and quality of polyphenols inhibiting enzymes. They proposed enzymatic methods to find out the fraction of inhibiting polyphenols in sorghum grain. The results of these methods correlated well with the DMF extractable polyphenols examined by ferric ammonium citrate in alkaline medium. In this chapter, the advantages of enzymatic and chemical methods to ascertain biologically active polyphenols were deliberated. Further, the author made some propositions to safeguard the brewing industry against unsuitable grain sorghum cultivars of the bird proof class (Klaus, 1975).

The association between sorghum grain polyphenol content, grain structure, and starch properties was determined. Starch from 10 sorghum varieties was isolated using an alkali steep and wet-milling procedure. White starch was obtained from SV2, a tannin-free variety with white pericarp. The sorghum varieties with red or white pericarp and high polyphenol levels provided pink starches. Correlation was not observed between the Hunter color values (L, a, b) of starches and grain polyphenol content. The appearance of grain in terms of pericarp color, or presence or absence of pigmented testa, had no relation with the intense pink color of starches. Starch amylose content was significantly and negatively correlated with grain floury endosperm texture. When compared with commercial maize starch, sorghum starches had higher peak viscosity (PV) in pasting. The sorghum starches took less time to reach PV from the initial viscosity rise when compared with maize starch. However, sorghum starches showed a higher rate of shear thinning (Rst) than maize starch. Grain polyphenol content was significantly and positively correlated with starch PV. Starch gel hardness exhibited a negative correlation with pasting properties of Rst and paste breakdown. Peak gelatinization temperature (Tp) appeared over a narrow range from 66 to 69°C. Tphad negative association with the floury endosperm part of the grain. The study specified that the sorghum starch properties are affected by grain polyphenol content and characteristics (Trust et al., 2001).

The modified acidified vanillin method of tannin analysis was used estimate relative tannin content of grain sorghum [*Sorghum bicolor* (L.) Moench]. The process includes overnight extraction of ground grain with methanol at room temperature. To the solution vanillin and hydrochloric acid an aliquot of the extract was added and the resulting color was read on a colorimeter at 500–525 mµ. The reagent is specific for astringent compounds in plants and the results were in correlation with digestibility (Robert, 1970).

Grain hardness influences different aspects of the growth and processing of cereal grain from resistance to fungal infection to cooking quality. Here, the biochemical basis and effects of hardness and grain strength in sorghum and maize were reviewed. It was found that the prolamins particularly γ-prolamins play a pivotal role in grain hardness. These prolamins shape the protein bodies and form disulfide bonds within themselves or with other proteins. The α-prolamins are the bricks and γ-prolamins form the cement. In hard grains and in the vitreous portion of hard grains prolamins are found in greater proportions. Further, the grain hardness is determined by the genetic and environmental effects on the quantities of the various prolamins and on their allocations within the protein body and in various sections of the endosperm. The deposition of prolamins and antifungal proteins (AFPs) is fast and higher in hard grains than soft grains. The hard and soft grains have different cell wall composition. The proportion of amylose is high in the starch of hard grain when compared to soft grain. The majority of the differences observed between hard and soft grains were seen between the outer and inner portion of the grain also. It is assumed that a master gene may control the onset of strength in grains by concurrently changing the levels of various apparently unrelated biochemical events. Further, they proposed that availability of solute may have a part in the regulation of expression of genes for hardness-related protein (Chandrashekar and Mazhar, 1999).

A study was carried out to determine the physicochemical characteristics of 45 sorghum genotypes. Significant variation was observed in the 100-grain weight, grain hardness, and grain water-soluble protein, amylose, and sugars contents of genotypes. The roti quality of flour from the 45 genotypes was evaluated in terms of flour color, appearance, taste, flavor, and texture by a trained taste panel. Instron machine was used to measure the dough texture. The associations between physicochemical characteristics of grain and roti qualities were studied. The results showed that the roti quality of the sorghum genotypes is mostly influenced by the quantity of water-soluble protein, amylose, and sugars in the grain (Subramanian and Jambunathan, 1981).

Sorghum is well adapted to hard environments and a staple food grain in many semi-arid and tropic areas of the world, particularly in Sub-Saharan Africa. The levels of starch (amylose and amylopectin) and starch depolymerizing enzymes are vital biochemical components for sorghum processing. To insure food security current research programs are focused on selecting varieties that can meet certain agricultural and food requirements from the great biodiversity of sorghums. Results showed that some sorghum cultivars were rich sources of micronutrients (minerals and vitamins) and macronutrients (carbohydrates, proteins, and fat). The presence of resistant starch (RS) in sorghum makes it interesting for obese and diabetic people. Additionally, sorghum may be an alternate food for people allergic to gluten. α-amylase and ß-amylase activities in malts of some sorghum varieties were similar to those of barley, which make them useful for a variety of agro-industrial foods. A specific emphasis was made on the effect of starch and starch degrading enzymes in the usage of sorghum for some African foods, e.g., "tô," thin porridges for infants, granulated foods "couscous," local beer "dolo" and agro-industrial foods such as lager beer and bread (Mamoudou et al., 2006).

The aroma impression of flavor volatile compounds present in sorghum malt beverage was evaluated. The samples were gathered and concentrated on a Tenax GC and identified on a GC-MS. In the sorghum malt beverage 28 volatile compounds were identified. These were made up of pyrazines, furans, aldehydes, ketones, esters, and alcohols. In the malt beverage volatiles the percentage of ketones, aldehydes, and esters was higher than the heterocyclic compounds and alcohols. The sorghum malt beverage had a characteristic nutty, sweet chocolate aroma which cannot be easily tied to a single volatile compound (Lasekan et al., 1997).

Though the condensed tannins are potentially significant antioxidants, it is generally believe that that tannin in sorghum has objectionable sensory characteristic. Rosemary et al. (2007) characterized the variances in the sensory characteristics of sorghums having different levels of total phenolic compounds were. The sensory characteristics of different sorghums were described and quantified by a trained sensory panel. All the sorghum cultivars were identified as bitter and astringent. The bran infusions of tannin free sorghums were perceived as sweeter and cloudy, whereas the infusions of tannin sorghums were perceived as darker, clearer, bitterer, and astringent. Sorghum whole-grain rice obtained from the tannin sorghums had comparatively soft endosperm texture (PAN 3860 and Ex Nola 97 GH) and perceived as dark, hard, chewy, bitter, and astringent. The grain from tannin free sorghums had hard endosperm texture (Segaolane and Phofu) and perceived as soft, sweet with a maize flavor. Unexpectedly, the bitterness and astringency, and other sensory characteristics of NS 5511 (tannin sorghum), were perceived as similar to PAN 8564 (tannin free sorghum), though the total phenol content of NS 5511 was more than those of PAN 8564. This suggested that objectionable sensory characteristics are not present in all condensed tannin containing sorghums.

11.3   FOOD PROCESSING

The literature on sorghum for human foods and on the association amongst some kernel characteristics and food quality was reviewed. The main foods prepared with sorghum like tortilla, porridge, couscous, and baked goods were described. It was found that the tortillas, made by mixing 75% of whole sorghum flour and 25% of yellow maize flour, are better than those prepared with whole sorghum flour only. A porridge prepared by mixing sorghum, maize, and cassava in 30:40:30 ratios were found to be most acceptable. Lower protein digestibility and higher biological value were recorded in the cooked porridge Aceda when compared to uncooked porridge Aceda. Though sorghum is not considered as bread making flour, the good quality bread can be produce by mixing 30% sorghum flour and 70% wheat flour (Anglani, 1998).

The effects of processing operations such as milling extraction, water soaking, malting, heat treatment and fermentation on phytate content of four Sudanese sorghum cultivars were analyzed. The conditions of processing utilized were: decortication to produce an 80% extraction meal; soaking in tap water for 12 and 24 h; 96 h germination; 3, 6, 9, and 12 h fermentation; and cooking at 95°C until starch gelatinized. Total phosphorus, phytate phosphorus and phytic acid were ascertained. The outcomes revealed that >85% of total phosphorus in sorghum cultivars was made up of phytic acid phosphorus. The phytic acid was reduced to various extents by all treatments. They found that the enzymatic methods remove phytic acid more effectively than physical extraction methods such as milling, soaking, and heating (Salah et al., 1998).

Sorghum is a gluten-free grain and can be used as an alternate to wheat flour. In this study, an effort was made to work out the association between sorghum grain quality factors and quality of Chinese egg noodles. Kernel characteristics, proximate analysis, and flour composition of four sorghum hybrids and end product of a Chinese egg noodle system were evaluated. The size of flour particle and the amount of starch damage were found to be affected by kernel size and weight. Noodles prepared from flour with fine particle size and high starch damage had high firmness and high tensile strength. Flour with smaller particle size (38 μm at 50% volume) and higher starch damage (6.14%) has shown highest water uptake. The cooking losses were less than 10% for all the samples. It was found that the starch of particle size <5 μm (C-type) gave a noodles with high firmness and tensile strength. Flour particle size, starch particle size and starch damage had a significant effect on water absorption. Therefore, by controlling the grain and flour quality characteristics of sorghum a Chinese egg noodle with good physical characteristics can be prepared (Liman et al., 2012).

Though both sorghum starch and sugar are feedstocks for production of biofuel, many studies on genetic improvement of sorghum concentrated on a single nonstructural carbohydrate, either grain starch or stem sugar. Therefore, in this study, genetic tradeoffs between sorghum grain and stem sugar was investigated. Twenty-seven traits associated with grain and stem sugar yield and composition were evaluated in a population obtained from sweet sorghum cultivar Rio and grain sorghum 'BTx623.' Total 129 quantitative trait loci (QTL) were detected across three environments. Tradeoffs detected between QTL of grain and stem sugar yield were colocalized with QTL of height and flowering time. Primarily, QTL were detected that improved yield and altered the composition of stem sugar and grain with no pleiotropic effects. For instance, a QTL on chromosome 3 that elucidated 25% of the genetic variance for stem sugar concentration did not colocalize with any grain QTL. These findings suggested that the selection of major QTL from both grain and sweet sorghum types could increase total nonstructural carbohydrate yield. The researchers concluded that the alteration of grain and stem sugar genetic potential for yield traits would bring about greater feedstock enhancement than alteration of composition traits (Seth et al., 2008).

Sorghum or bread wheat flours were composited with defatted soy flour to produce biscuits. Double protein content and 500–700% more lysine content was found in sorghum-soy and bread wheat-soy 1:1 ratio composite biscuits, when compared to the 100% cereal biscuits. In the sorghum-soy biscuits 170% increased IVPD was observed. It was found that 3 to 10 year-olds can get 50% of the recommended daily protein intake from two such biscuits of 28 g. Descriptive sensory assessment unveiled that sorghum-soy composite biscuits had crispy and dry texture characteristics. Sensory assessment by schoolchildren has shown that the composite biscuits were rated as acceptable as the cereal only biscuits, and this was continued over 4 days of evaluation. Therefore, sorghum-and bread wheat-soy biscuits have great potential as protein-rich supplementary foods to reduce protein energy malnutrition in children (Charlotte et al., 2010).

Patties were made with 20% fat ground beef and sorghum flour (SF) at 2, 4, and 6% levels (10, 20, and 30% as rehydrated 1:4 with water). The addition of SF increased pH, yield, and reduced total cooking loss, shrinkage in diameter, and increase in thickness in beef patties. However, it did not show any effect on water activity of cooked patties. With the increase in the SF level an increment in fat and water retentions of beef patties was noted. The increase in level of SF reduced Hunterlab a values (redness) for raw patties, however, it did not influence the values for cooked patties. The increased level of SF reduced the shear force and compression of cooked patties. The reduction in meat aroma and flavor, but increment in sorghum aroma and flavor of cooked patties was observed at higher levels of SF. The increased SF level increased the tenderness of cooked patties, without any effect on juiciness of cooked patties (Jen-Chieh et al., 1999).

The aroma, crumb flavor, top crust, flavor, and texture characteristics of the sorghum bread were compare with a commercial rye bread by conducting a descriptive test. Bread was prepared from 50% sorghum based composite flour and tested by six trained panellists. The panellists associated with sorghum observed hay like attribute in the analysis of aroma. In the crumb and top crust flavors slight sourness and astringency were experienced. Due to the variations in formulation of the rye and sorghum flours, textural variances existed between the two breads. The acceptability of sorghum composite bread's with 37 consumers lead to an overall acceptability mean score of 6.9 on a nine-point hedonic scale (Lin et al., 2001).

Tortilla chips were prepared by processing white maize (*Zea mays* L.), tan plant color white sorghum (*Sorghum bicolor* Moench) and a variety of blends of these in a pilot plant. The most favorable conditions for making of tortilla chips were: (1) lime-cooking at boiling temperature for 60 and 20 min with 1 and 0·5% lime, respectively;, (2) quenching to drop the temperature of the cooking liquor to 68°C; (3) steeping the grains for 8–12 h and 4–6 h, respectively. The resultant nixtamals (lime-cooked and steeped grains) were stone-ground into coarse dough (masa). The masa was sheeted, cut into triangles, and baked for 39 sec at 280°C into tortilla pieces, which were fried for 1 min at 190°C to make tortilla chips. During these conditions, more dry matter (DM) losses were observed in maize nixtamal compared to sorghum nixtamal. The machinability and moisture content were similar in both grain masas. An increment in insoluble dietary fiber was noted with the increase in sorghum levels. All chips showed similar oil uptake during frying. Panelists observed a significant difference in bland taste of sorghum chips and characteristic taste of lime-cooked maize. Tortilla chips having 50% maize had a somewhat lower ($P < 0.05$) flavor, texture, and aroma than maize tortilla chips. Thus, snack products in which a strong maize flavor is not preferred could be produced from sorghum. Up to 50% sorghum blends with maize would give acceptable products (Serna-Saldivar et al., 1988).

The study focused on assays of odorant retention from an essential oil mixture by native corn, sorghum, and amaranth starches using capillary gas chromatography (GC) and differential scanning calorimetry (DSC). There are 30 main compounds in essential oil mixture, which include monoterpene and sesquiterpene hydrocarbons, alcohols, ketone, phenols, and ester. The chemical composition (amylose, protein, and surface lipids content), surface properties (surface area, pore diameter), and microstructure (SEM) of native starches were characterized. The starches were stored at room temperature in the dark for 2 days (reference sample), 3 and 7 months after mixing odorants and stirring intensively. It was found that the retention of aroma compounds was affected by the chemical properties of odorants, their composition in the mixture along with starch surface properties upon storage. During storage, a considerable loss in monoterpene hydrocarbons was observed irrespective of starch botanical origin. Amaranth and sorghum starches showed a higher retention of alcohols, ketone, phenols, and sesquiterpene hydrocarbons after 3 months of storage. However, their retention diminished after long storage, particularly in the case of amaranth starch. The DSC results of stored corn and sorghum starches with odorants revealed an extra endothermic contribution, which showed that the gelatinized starch matrixes retains more aroma compounds. Additionally, they found that the odorants interacts with the solubilized amylose (Wioletta et al., 2013).

A Nigerian fermented porridge, ogi was prepared by processing seven sorghums. To find out which sorghums had the best properties for ogi preparation, the color, taste, texture, aroma, and consistency of the ogis were estimated. In general, the highest ratings for these properties were found in nonwaxy, white sorghums ogi. Ogi prepared from waxy sorghums had poor and undesirable consistency. The brown high tannin sorghums produced ogi with undesirable brownish red color, poor consistency and texture, and low *in vitro* starch and protein digestibility. Ogi prepared with TAM680 and Funk G776W resembled the Nigerian ogi (Akingbala et al., 1981).

Sorghum is cereal which can be cultivated efficiently in the semi-arid regions of the world. Though sorghum has been used conventionally to make foods, malt beverages and beer, its structure and function have not been studied to the same level as grains which are cultivated in the more developed areas of the world. Similar to other cereals, sorghum is an outstanding source of starch and protein. It can be processed into starch, flour, grits, and flakes and utilized to produce a variety of industrial products. Further, sorghum can be malted and processed into malted foods, beverages, and beer. If relevant scientific work is undertaken to increase the yield, quality, and knowledge needed by locally based industries, sorghum could play a significant role in development of agriculture in some of the poorest countries of the world (Palmer, 1992).

Sorghum and millets are gluten free and have great potential in foods and beverages. They are suitable for coeliac. Sorghum is a vital source of nutraceuticals, antioxidant, phenolics, and cholesterol-lowering waxes. Sorghum and sometimes millets can be used to produce cakes, cookies, pasta, a parboiled rice-like product, and snack foods. Wheat-free sorghum or millet bread remains the main challenge. The additives such as native and pre-gelatinized starches, hydrocolloids, fat, egg, and rye pentosans could improve the quality of sorghum bread. However, sorghum bread has low specific volumes and tends to stale faster when compared to wheat bread or gluten-free breads. Lager and stout beers with sorghum are brewed commercially. The complete substitution of barley malt with sorghum malt is restricted due to the high-starch gelatinization temperature and low beta-amylase activity of sorghum. Millets brewing is still at an investigational stage. Sorghum could be important for bioethanol and other bio-industrial products. Bioethanol research has concentrated on increasing the economics of the process through variety selection, development of method for low-quality grain and pre-processing to get valuable byproducts. Due to the hydrophobicity of sorghum byproducts such as kafirin, prolamin proteins, and the pericarp wax, they have potential to be used as bioplastic films and coatings for foods (John et al., 2006).

Starch pasting properties and protein heat damage of whole-grain meals from barley, millet, rye, and sorghum separately and in mixture with wheat flour were evaluated during cycles of heating and cooling in RVA tests. Bread, cake, cookie, or snack products were prepared by mixing whole-grain meals with either hard or soft wheat flour. Then physical properties and acceptability of these products were evaluated. Between cereals significant differences were noted in starch peak, breakdown, setback viscosities and in protein PV. The outcomes unveiled that cereal blends could be formulated by using RVA with some pasting properties. Replacement of wheat flour, with 15% of barley, rye, millet or sorghum whole-grain, had no major harmful effects on physical properties or acceptability of pita bread. Furthermore, substitution of wheat flour with up to 30% of barley, rye, millet or sorghum whole-grain meal did not show any noteworthy effects on cakes or cookies quality. Further, they developed a multigrain snack-like food as a healthy product and were highly acceptable in a sensory test. This developed product would increase consumption of whole-grain foods, resulting in increased intake of fiber and health-enhancing components (Sanaa et al., 2004).

The major advances that have been made in the development of sorghum food processing technologies were described in this chapter. Further, they examined the challenges that still have to be met. Several advances have been made in developing sorghum food processing technologies. Two instances where the technologies have been effectively implemented are: the industrialization of sorghum brewing, taking it from a rural craft to a 20,000-liter batch scale, and brewing conventional beer with sorghum grain and enzymes. However, the application of other technologies, such as composite breads, is notably lacking. The unavailability of cost-efficient, dependable supplies of sorghum grain of acceptable quality for preparing high quality flour was found to be a major factor that limited the use of sorghum in Africa. The challenge, therefore, seems to revolve around a holistic approach to implementation, involving: economic studies, government programs, seed supply, grain production, selection of suitable technologies, and training of operators, consumer awareness, and grain and product quality standards (John and Janice, 2001).

11.4 RESEARCH ADVANCES

People with celiac disease have an autoimmune reaction to gluten proteins present in wheat and its related cereals like barley and rye. This reaction forms autoantibodies and damages the villi in the small intestine, which results in malabsorption of nutrients and other gluten-induced autoimmune diseases. Sorghum can be potentially developed into an important crop for human food products. The flour of white sorghum hybrids is light in color with a bland, neutral taste that does not provide unusual colors or flavors to food products. These qualities make it attractive for use in wheat-free food products. Though sorghum is considered as a safe food for celiac patients, mainly because of its association to maize, direct testing has not been performed on its safety for gluten intolerance. So, research works are required to evaluate its safety and tolerability in celiac patients. In this study, safety and tolerability of sorghum flour products was assessed in adult celiac disease patients, using an *in vitro* and *in vivo* challenge. Sorghum protein digests did not cause any morphometric or immune mediated alteration of duodenal explants from celiac patients. Patients fed daily for 5 days with food product obtained from sorghum had no gastrointestinal or non-gastrointestinal symptoms and the level of anti-transglutaminase antibodies remained same at the end of the 5-days challenge. The study proposed that products obtained from sorghum had no toxic effects on celiac patients in both *in vitro* and *in vivo* challenge. Hence sorghum can be a safe food for people with celiac disease (Carolina et al., 2007).

QTL for grain quality, yield components, and other traits were examined in two sorghum caudatum×guinea recombinant inbred line (RIL) populations. Total 16 traits viz., plant height, panicle length, panicle compactness, number of kernels/panicle, thousand-kernel weight, kernel weight/panicle, threshing percentage, dehulling yield, kernel flouriness, kernel friability, kernel hardness, amylose content, protein content, lipid content, germination rate and molds during germination and after harvest were evaluated. These traits associated to two 113- and 100-point base genetic maps by means of simple (SIM) and composite (CIM) interval mapping. In both populations the number, effects, and relative location of identified QTLs were found to be compatible with the distributions, heritabilities, and associations among traits. Some chromosomal segments had a marked effect on several traits and were expected to harbor major genes. The locations of these QTLs were discussed with respect to the previous studies on sorghum and other grasses. Based on their relative effects and location, these QTLs, could be used as targets for marker-assisted selection (MAS) and offer an opportunity to accelerate breeding programs (Rami et al., 1998).

A study was undertaken to assess breeding potential of the sorghum population. Two S_2 or S_1-families obtained from each of 118 S_0 plants of randomly mated sorghum population (PP18) were evaluated. Genetic parameters such as days to bloom, plant height, lodging, panicle compactness, panicle weight, kernel weight, weathering, hardness, and vitreosity were estimated. Weighted least-squares procedure was used to estimate additive variances. Significant additive variances were observed for all traits. The additive genetic coefficients of variation varied from 2.9% for days to bloom to 25.5% for lodging. In S_0 derived family selection narrow-sense heritabilities ranged from 0.349 to 0.734, while for S_1-derived family selection it ranged from 0.359 to 0.842. At both genotypic and phenotypic levels high correlations were observed among agronomic traits and quality traits. However, the correlation between agronomic and quality traits was low. Kernel hardness was negatively correlated with weathering and positively correlated with vitreosity. For S_0- and S_1-derived family selection direct selection responses (percentage of mean) ranged from 3.70 to 28.58% and from 5.10 to 37.10%, respectively. Higher direct responses were recorded for most of the traits. However, there was no significant difference between direct and correlated response for some pairs of the traits. The study findings put forward that all the tested traits have good potential for conventional breeding methods and early generation selection. A further improvement would be expected from S_1-derived than from S_0-derived family selection (Osman et al., 1985).

In sorghum, most important storage proteins are kafirins and they forms protein bodies with poor digestibility. The essential amino acid lysine is absent in these proteins and impart poor protein quality to the kernel. Major portion of the total kafirins was made up of α-kafirins, which are encoded by the *k1C* family of highly similar genes. In this study, the *k1C* genes were targeted by a clustered regularly interspaced short palindromic repeats (CRISPR)/CRISPR-associated protein 9 (Cas9) gene-editing approach to develop varieties with high kafirin levels and improved protein quality and digestibility. They designed a single guide RNA (sgRNA) to cause mutations in a conserved region encoding the endoplasmic reticulum signal peptide of α-kafirins. A sequencing of kafirin PCR products revealed large editions in 25 of 26 events in one or multiple *k1C* family members. Reductions in α-kafirin levels were recorded in T_1 and T_2 seeds. The selected T_2 seeds exhibited considerable increase in grain protein digestibility and lysine content. Therefore, a single consensus sgRNA carrying target sequence mismatches is enough for wide editing of all *k1C* genes. The resultant quality enhancements can be deployed rapidly for breeding to develop transgene-free, improved sorghum cultivars (Aixia et al., 2018).

11.5   CONCLUSIONS

To enhance sorghum grain quality, the grain quality characteristics should be determined. In this chapter, various methods that can be utilized to determine the physical and chemical characteristics of sorghum grain quality are summarized. The processing methods to improve grain quality and to make the best use of sorghum byproducts are also presented. Further, the recent progresses made in sorghum grain quality research are discussed.

KEYWORDS

- **diastatic power**
- **food quality**
- **grain quality**
- **peak viscosity**
- **single kernel characterization system**
- **sorghum**

REFERENCES

Aixia, L., Shangang, J., Abou, Y., Zhengxiang, G., Shirley, J. S., Chi, Z., Ruthie, A., Thomas, E. C., & David, R. H., (2018). Editing of an alpha-kafirin gene family increases, digestibility, and protein quality in sorghum. *Plant Physiology Journal, 6*, 15–22.

Akingbala, J. O., Rooney, L. W., & Faubion, J. M., (1981). Physical, chemical, and sensory evaluation of ogi from sorghum of differing kernel characteristics. *Journal of Food Science, 12*, 45–54.

Anglani, C., (1998). Sorghum for human food: A review. *Plant Foods for Human Nutrition, 52*, 85–95.

Aruna, C., & Audilakshmi, S., (2004). Genetic architecture of grain hardness: A durable resistance mechanism for grain molds in sorghum [*Sorghum bicolor* (L.) Moench]. *The Indian Journal of Genetics and Plant Breeding, 64*, 35–38.

Audilakshmia, S., Arunaa, C., Solunkeb, R. B., Kamatarc, M. Y., Kandalkard, P., Gaikwade, K., Ganesh M. K., et al., (2007). Approaches to grain quality improvement in rainy season sorghum in India. *Crop Protection, 26*, 630–641.

Bean, S. R., Chung, O. K., Tuinstra, M. R., Pedersen, J. F., & Erpelding, J., (2006). Evaluation of the single kernel characterization system (SKCS) for measurement of sorghum grain attributes. *Cereal Chemistry, 10*, 83–95.

Betaa, T., Rooneyb, W. L., Marovatsangaa, L. T., & Taylorc, J. R. N., (2000). Effect of chemical treatments on polyphenols and malt quality in sorghum. *Journal of Cereal Science, 31*, 295–302.

Carolina, C., Luigi, M., Nicola, C., Cristina, B., Luigi, D., Giudiced, D., Rita, M., et al., (2007). Celiac disease: *In vitro* and *in vivo* safety and palatability of wheat-free, sorghum food products. *Clinical Nutrition, 26*, 799–805.

Chandrashekar, A., & Mazhar, H., (1999). The biochemical basis and implications of grain strength in sorghum and maize. *Journal of Cereal Science, 30*, 193–207.

Charlotte, A. S., Henriëtte, L. D. K., & John, R. N. T., (2010). Nutritional quality, sensory quality, and consumer acceptability of sorghum and bread wheat biscuits fortified with defatted soy flour. *International Journal of Food Science and Technology, 28*, 1–10.

De Morais, C., Leandro, P., Soraia, S. M., Hércia, S. D., Pinheiro-Sant'Ana, & Helena, M., (2017). Sorghum (*Sorghum bicolor* L.): Nutrients, bioactive compounds, and potential impact on human health. *Critical Reviews in Food Science and Nutrition, 6*, 234–243.

Duodua, K. G., Nunesb, I. A., Delgadillob, M. L., Parkerc, E., Millsc, N. C., & Beltonc, P. S., (2002). Effect of grain structure and cooking on sorghum and maize *in vitro* protein digestibility. *Journal of Cereal Science, 35*, 161–174.

Eggum, B. O., Monowar, L., Bach, K. K. E., Munck, L., & Axtell, J., (1983). Nutritional quality of sorghum and sorghum foods from Sudan. *Journal of Cereal Science, 1*, 127–137.

Elemo, G. N., Elemo, B. O., & Okafor, J. N. C., (2011). Preparation and nutritional composition of a weaning food formulated from germinated sorghum *(Sorghum bicolor)* and steamed cooked cowpea (*Vigna unguiculata* Walp.). *American Journal of Food Technology, 6,* 413–421.

Jen, C. H., Joseph, F. Z., & Jane, A. B., (1999). Functional properties of sorghum flour as an extender in ground beef patties. *Journal of Food Quality, 22*, 51–62.

John, R. N. T., & Janice, D., (2001). Developments in sorghum food technologies. *Advances in Food and Nutrition Research, 43,* 217–264.

John, R. N. T., Tilman, J. S., & Scott, R. B., (2006). Novel food and non-food uses for sorghum and millets. *Journal of Cereal Science, 44*, 252–271.

Joshua, H. W., Tsang, L. N., Caia, J. S., Jeffrey, F. P., William, H. V., William, J. H., Jeff, D. W., & Peggy, G. L., (2009). Digestibility of protein and starch from sorghum (*Sorghum bicolor*) is linked to biochemical and structural features of grain endosperm. *Journal of Cereal Science, 49*, 73–82.

Klaus, H. D., (1975). Enzyme inhibition by polyphenols of sorghum grain and malt. *Journal of the Sciences of Food and Agriculture, 26*, 1399–1408.

Lasekan, O. O., Lasekan, W. O., & Idowu, M. A., (1997). Flavor volatiles of 'malt beverage' from roasted sorghum. *Food Chemistry, 58*, 341–344.

Liman, L. T. J., Herald, D. W., Jeff, D. W., Scott, R. B., & Fadi, M. A., (2012). Characterization of sorghum grain and evaluation of sorghum flour in a Chinese egg noodle system. *Journal of Cereal Science, 55*, 31–36.

Lin, C., Carole, S., & Xiuzhi, S. S., (2001). Sensory characteristics of sorghum composite bread. *Food Science and Technology, 35*, 465–471.

Mamoudou, H. D., Harry, G., Alfred, S. T., Alphons, G., Voragen, J., & Willem, J. H., (2006). Sorghum grain as human food in Africa: Relevance of content of starch and amylase activities. *African Journal of Biotechnology, 5*, 20–29.

Osman, E. I., Nyquist, W. E., & Axtell, J. D., (1985). Quantitative inheritance and correlations of agronomic and grain quality traits of sorghum. *Crop Science, 25*, 649–654.

Palmer, G. H., (1992). Sorghum-food, beverage and brewing potentials. *Process Biochemistry, 27*, 145–153.

Rameshwar, S., & John, D. A., (1973). High lysine mutant gene (hl) that improves protein quality and biological value of grain sorghum. *Crop Science, 13*, 535–539.

Rami, J. F., Dufour, P., Trouche, G., Fliedel, G., Mestres, C., Davrieux, F., & Blanchard, P., (1998). Quantitative trait loci for grain quality, productivity, morphological and agronomical traits in sorghum (*Sorghum bicolor* L. Moench). *Theoretical and Applied Genetics, 97*, 605–616.

Robert, E. B., (1970). Method for estimation of tannin in grain sorghum. *Agronomy Journal, 63*, 511–512.

Rooney, L. W., & Murty, D. S., (1981). Evaluation of sorghum food quality. In: *Sorghum in the Eighties: Proceedings of the International Symposium on Sorghum*. Patancheru. A.P. India.

Rosemary, I., Kobue, L., & John, R. N. T., (2007). Effects of phenolics in sorghum grain on its bitterness, astringency and other sensory properties. *Journal of the Science of Food and Agriculture, 87,* 1940–1948.

Salah, E. O., Mahgoub, S., & Elhag, A., (1998). Effect of milling, soaking, malting, heat-treatment, and fermentation on phytate level of four Sudanese sorghum cultivars. *Food Chemistry, 61*, 77–80.

Sanaa, R., El-Sayed, M., & Abdel-Aalb, (2006). Pasting properties of starch and protein in selected cereals and quality of their food products. *Food Chemistry, 95*, 9–18.

Sergio, O. S., Saldivar, J. E., & Ramírez, (2019). Grain structure and grain chemical composition. *Sorghum and Millets: Chemistry, Technology, and Nutritional Attributes* (2nd edn., pp. 85–129).

Serna-Saldivar, S. O., Tellez-Giron, A., & Rooney, L. W., (1988). Production of tortilla chips from sorghum and maize. *Journal of Cereal Science, 8*, 275–284.

Seth, C. M., Arun, S., William, L. R., Patricia, E. K., John, E. M., Sharon, E. M., & Stephen, K., (2008). Genetic improvement of sorghum as a biofuel feedstock: I. QTL for stem sugar and grain nonstructural carbohydrates. *Crop Science, 48*, 2165–21791.

Subramanian, V., & Jambunathan, R., (1981). Properties of sorghum grain and their relationship to roti quality. In: *Proceedings of the International Symposium on Sorghum Grain Quality*. ICRISAT Patancheru.

Trust, B., Harold, C., Lloyd, W. R., & John, R. N. T., (2001). Starch properties as affected by sorghum grain chemistry. *Journal of the Sciences of Food and Agriculture, 81*, 245–251.

Wioletta, B., Tamara, A., Misharina, D., Fessasc, M. S., & Adrian, R. G., (2013). Retention of aroma compounds by corn, sorghum, and amaranth starches. *Food Research International, 54*, 338–344.

CHAPTER 12

Research Advances in Breeding and Biotechnology of Sorghum

ABSTRACT

As sorghum is an economically and socially significant crop, several advances have been made in its research. Increased applications of new biotechnological methods could bring about an increased rate of improved sorghum production. Here, the research progresses made in sorghum breeding, and biotechnology are reviewed.

12.1 GENETIC DIVERSITY (GD) OF SORGHUM GERMPLASM COLLECTION

Almost a century has been spent gathering and conserving plant genetic diversity (GD). To improve agricultural crops, germplasm banks that serve as sources of genetic variation and different genes have been established. Further, the construction of genetic linkage maps have made the study of chromosomal locations of genes possible, which helps in improving yield and other agriculturally important traits. Finally, the genome research tools may release the genetic potential of our wild and cultivated germplasm resources for the benefit of society.

Ethiopia is one of the centers of origin for sorghum. The GD study from its center of origin is vital for its enhancement. So, in this study inter simple sequence repeat (SSR) markers were used to assess the within and among population GD of sorghum from Ethiopia. DNA was extracted from young leaves of sorghum and PCR was performed with seven primers. A total of 55 clear and reproducible bands with 100% polymorphism were produced. The total GD was 0.205 and Shannon's diversity index (I) was 0.296. The highest gene diversity (0.3) was observed in population from North

Gondar Intermediate altitude and North Shewa population had lowest diversity (0.1). AMOVA revealed that within populations contributed 44.45% of the total variation, followed by among populations (30.84%) and among groups (24.71%). Group based UPGMA showed four separate clusters, while the population-based UPGMA indicated that all populations grouped within its own ecological zone excluding North Gondar (I1) intermediate altitude population and improved varieties (IVs) and stay-green (IM) individuals that were grouped in the highland populations. Highest GD was recorded in intermediate altitude populations indicating that the high and low altitude populations might have originated from the intermediates through ecological adaptation (Hailekiros and Tileye, 2013).

This was the first study on patterns of GD of sorghum landraces at the local scale. Knowledge on landrace diversity helps in interpretation of evolutionary forces under domestication, and it has uses in the conservation and use of genetic resources in breeding programs. Fifty-nine named sorghum taxa, representing 46 landraces were distinguished by Duupa farmers in a village in Northern Cameroon. To provide the potential for extensive gene flow seeds were sown as a mixture of landraces (mean of 12 landraces per field) in each field. Spatial patterns of planting and farmers' perceptions of landraces were recorded and SSR markers were used to characterize 21 landraces. Distance and clustering methods analysis were done and 21 landraces were grouped into four clusters. These clusters corresponded to functionally and ecologically diverse groups of landraces. Within-landrace 30% of total variation was contributed by genetic factors. The average F_{st} over landraces was 0.68, indicating high inbreeding within landraces. There was substantial and significant difference among landraces. It was found that patterns of genetic variation were affected by the historical factors, difference in breeding systems, and farmers' practices. Farmers' practices were important in the maintenance, in spite of gene flow, of landraces with different combinations of agronomically and ecologically relevant traits. Therefore, they must be considered in approaches of genetic resources conservation and use (Adeline et al., 2007).

The appropriateness of two DNA-based fingerprinting methods, SSRs and random amplified polymorphic DNA (RAPD) analyses for quantifying GD was compared. Twenty-two sorghum genotypes from an array of germplasm sources with agronomically important traits were selected and screened for polymorphism with 32 RAPD primers and 28 sets of sorghum SSR primers. The results showed that SSR markers were highly polymorphic with an average of 4.5 alleles per primer. The RAPD primers were less polymorphic with almost 40% of the fragments being monomorphic. The GD among sorghum lines was also examined. Strong correlation was detected between the genetic distances computed from SSR and the distances based on the geographic origin and race classifications. Therefore, the study results concluded that SSR markers seem to be mainly useful for the assessment of genetic correspondence among diverse genotypes of sorghum (Agrama and Tuinst, 2003).

The diversity in 28 Eritrean sorghum landraces was characterized and compared with the representative samples of the world sorghum collection using a precise high-throughput method. Group of SSR markers were sized and scored on automated DNA-sizing gels. When compared to other sorghum germplasms, very high level of diversity in terms of both number and size range of SSR alleles was observed among the 28 Eritrean landraces. A high level of within-population diversity was observed in individual landraces. Heterozygosity and between-population diversity was also high. Most of the (but not all) classifications could group the Eritrean sorghums into 7–10 major subgroups which are in agreement with descriptions by farmers. However, most Eritrean sorghums grouped in a separate subgroup, eight of the Eritrean landraces especially those from Ethiopia/Sudan and India or of the durra and caudatum races were clustered with other sorghums in the world collection. These findings indicated that a huge amount of germplasm diversity and genetic novelty are available in Eritrean sorghums, and that SSR markers could help in wise use of this diversity for sorghum research and improvement (Ghebru et al., 2002).

12.2 HIGH-YIELDING VARIETIES

Renewable energy is an important source of energy that adds to energy security, decreasing dependency on fossil fuels and emission of greenhouse gases. India requires more than 6.3 billion liters of ethanol. Sweet sorghum is a promising dry land adapted biofuel feedstock that tackles food-versus-fuel issue advantageously. Because of its genetic variability relating to stalk sugar traits such as total soluble sugars, green stalk yield, juice quantity, and grain yield several research institutes in India and abroad have developed IVs and hybrids. Two commercial sweet sorghum based distilleries were founded in India but could not operate for long due to some reasons. The decentralized crushing units were formed to over-whelm the issues met by centralized units. The availability of improved cultivars with higher sugar yield and multiple biotic and abiotic stress tolerance in addition to the policy support from government of India in terms of both producer and processor incentives materialize could help in the large-scale cultivation of sweet sorghum (Vinutha et al., 2014).

The performance of local and improved sorghum varieties was deter-mined by conducting an experiment in central zone (CZ), Tanzania. One local variety (udo) and two IVs (NACO-1 and macia) were selected for the study. Among tested varieties significant variation was noted for plant height, leaf area, and leaf length, whereas there was no significant varia-tion in terms of yield, number of tillers, leaf width, total leaf area, leaf area index (LAI), and moisture content. The highest yield was obtained from improved variety NACO-1 (1870.83 g) followed by macia (1412.7 g) whereas the yield of local variety udo was lowest (914.43 g). The results indicated the presence of difference in genetic bases with variation in the gene action that expresses phenotypes of tested varieties. Among the tested varieties, the QTLs expressing high yielding were detected in the variety NACO-1. This may be the reason for high yielding ability of NACO-1 variety. Therefore, this study recommended the variety NACO-1 for adop-tion by farmers to increase productivity of sorghum, as this variety has performed well in most sorghum production areas in Tanzania (Andekelile and Zacharia, 2018).

The stability parameters are defined by the model, $Y_{ij} = \mu_1 + \beta_1 I_j + \delta_{ij}$. This model could be employed to describe the performance of a variety across different environments. Y_{ij} is the mean of the ith variety at the j_{th} environment, μ_1 is the mean of ith variety across all environments, β_1 is the regression coefficient that measures the response of the ith variety to changing environments, δ_{ij} is the deviation from regression of the ith variety at the jth environment, and I_j is the environmental index. To find out whether genetic differences could be identified by this model, the data from two single-cross diallels and a set of 3-way crosses was studied. Among lines, genetic differences were specified by the regression of the lines on the environmental index without any indication of nonadditive gene action. Many hybrids showed near zero estimates of the squared deviations from regression, while other hybrids had very large estimates (Eberhart and Russell, 1996).

Sorghum is a staple food of people residing in the subtropical and semiarid regions of the world. Sorghum is cultivated under diverse agro ecosystems around the world. The introduction of improved cultivars, higher input use, and effective crop management has been successfully improved the sorghum grain yield over the years. Improved yield and adaptation are the foremost purposes of every sorghum-breeding program. In sorghum wide GD is available, which is exhibited in its five basic races, viz., bicolor, kafir, guinea, caudatum, and durra and their 10 intermediate races. The Zera-zera (an intermediary race between caudatum and guinea) landraces from Ethiopia and Sudan have proved to be valuable sources for many characteristics such as exceptional grain quality, high grain yield potential, tan plant, resistance to leaf diseases, and desired plant type. In spite of the substantial diversity in the available germplasm, very few germplasm lines have been exploited so far. Therefore, to develop improved sorghum cultivars for sustainable production its genetic base should be broaden by using the diversity among the five basic races (Aruna et al., 2019).

The effect of variety characteristics on adoption of improved sorghum varieties in Kenya was investigated. The variety specific drivers of adoption were recognized in a random sample of 140 farmers using mideultivariate probit. The results of farmer's variety characteristics perception revealed that IVs had desired production and marketing characteristics, though the local varieties were supposed to have the best consumption characteristics. Further evidences showed that the taste, drought tolerance, yield, ease of cooking, and the ability of variety to get a price premium are the important sorghum variety characteristics that drive rapid adoption. It was found that the early maturity which is a main focus of research have no effect on the adoption decision. Therefore, the study results suggested that the breeders should also concentrate on non-yield characteristics like taste and ease of cooking while developing improved seed varieties, so as to increase adoption and meet the various needs of the farmers (Anne et al., 2014).

The major sorghum ecological zones in Ethiopia are high altitudes, intermediate altitudes with high rainfall, low altitudes with low rainfall, and low altitudes with high rainfall. To develop varieties and/or hybrids for all of these zones, the strategies used in the national sorghum improvement program were described. The techniques used and described as examples were germplasm collection and introduction, backcross breeding, pedigree breeding, and hybrid breeding. Where the availability of high-level trained manpower and resources are limited, the approaches described maximize efficient use of trained manpower and reduce the time per unit of plant breeding advance. The procedures used in coordination and centralization of key activities in the national sorghum breeding program were described. It is suggested that such a coordinated and centralized breeding approach be used in the national breeding programs of the other major crops of the country. Sorghum is grown under diverse environmental conditions in Ethiopia. It is the dominant crop in the lowlands of the country, where rainfall is limited. It is also widely grown under high rainfall lowland situations. In some parts of the country, it is cultivated as the major crop in the highlands also. The national sorghum-breeding program must be geared towards fulfilling the improved variety and/or hybrid sorghum needs of these Zones (Brhane, 1981).

Fodder sorghum CO 31 was developed and released from Coimbatore in 2014. It is a Gamma-ray (400 Gy) mutant of multicut fodder sorghum variety CO (FS) 29. In the field trails CO 31 recorded an average green fodder yield of 201 t/ha/yr, whereas the yield of CO (FS) 29 was 161.2 t/ha/yr. The green fodder yield of CO 31 was 16.3% higher than CO (FS) 29. After flowering the HCN content in CO 31 was 172 ppm, whereas CO (FS) 29 had 185 ppm HCN which reduces the risk of HCN toxicity to the animals. Further, CO 31 had less fiber content (19.80%) that gives increased digestibility and palatability. After physiological maturity, the seeds in CO 31 were more intact than CO (FS) 29. Furthermore, the non-shattering behavior of CO 31 seeds enables improved seed yield to the tune of 19.1% than CO (FS) 29 (Iyanar et al., 2015).

The performance and moisture stress adaptability of seven sorghum varieties was tested in Fedis and Babile districts of Eastern Hararghe. The sorghum varieties viz. Melkam, Meko, Teshale, Gambela 1107, Dhaqaba, Macia, and Birhan were used in the study. The high significant difference was observed interms of parameters like days to 50% flowering, days to maturity, plant height and panicle length. The earliest days to 50%flowering were observed in Meko variety with 68.67 days at Fedis and 66.67 days at Babile. The Dhaqaba variety exhibited late flowering with 78.33 and 76.67 days at Fedis and Babile, respectively. Further, Meko variety showed earliest days to maturity by taking 110.3 and 106.3 days at Fedis and Babile, respectively. The highest grain yield was recorded in Melkam (6122 kg ha^{-1}) variety followed by Macia (5751 kg ha^{-1}), Gambela 1107

(5700 kg ha^{-1}) and Dhaqaba (5537 kg ha^{-1}) at Babile. Simultaneously, the lowest yield of 3759 and 4255 kg ha^{-1} was achieved from Meko and Birhan, respectively. Therefore, the study recommended these varieties for further evaluation under farmer's management (Fuad et al., 2018).

Sorghum is extensively cultivated for food, feed, fodder, and fuel in the semi-arid tropics of Asia, Africa, the Americas, and Australia. However, the competition from other remunerative crops rapidly declined the area under sorghum cultivation in Asia, its grain production levels have not dropped at the same rate due to the adoption of high yielding hybrids. Biotic and abiotic stresses, for example, shoot fly, stem borer, grain molds, and terminal drought stress remained as foremost challenges in improving sorghum productivity. International Crops Research Institute for the Semi-Arid Tropics (ICRISAT) and the respective national programs are working on genetic improvement of sorghum for high yield; shoot fly, and grain mold resistance, and sweet stalk traits. Other research focuses of ICRISAT are adaptation to post rainy season, terminal drought tolerance, and improving micronutrient contents (Fe and Zn) in sorghum grain. Genetic and cytoplasmic variation of hybrid parents and varieties for important traits is essential for sustaining the productivity gains. To enhance the market value of sorghum, particular attention should be given to the grain and stover quality needs of different market segments in sorghum improvement research (Kumar et al., 2011).

Participatory plant breeding (PPB) strive for the involvement of farmers more closely in crop improvement to improve breeding influence. Though PPB intends to improve breeding practice, there has been little analysis of the recent practice breeding institutions. Such a study is essen-tial, both to empathize why a breeding program works the way it does, and to assess the potentials of improvements. In this chapter, theories of path-dependency, social construction of technology, and actor-networks were developed to analyze the historic development of the Ethiopian Sorghum Improvement Program (ESIP), a long-running and high-level public-sector attempt whose outputs had limited adoption. This analysis investigates options in technology development, the social networks affecting them, and the probability that recognized options become stabi-lized in a pathway that resists deviations to another lines of research and technology development. Application of this analysis to ESIP assists in understanding the path-dependency of sorghum breeding, indicating how principal choices around agroecological classifications, germplasm use, and F_1 hybrid development became 'locked-in,' thus resisting change. Technical limitations, breeding practices, and actor systems all strengthen specific choices from the past, as does the centralized organization of the ESIP team. Most PPB efforts think that poor breeder understanding of the traits desired by farmers is the foremost reason for low impact, and thus this gap should be addressed. This study points out major reasons for poor impact, and reveals that institutional change in breeding is not likely to come out from a PPB intervention concentrating on selection criteria alone. In order to be permanent, improvements need to identify technical pathways, reinforce the voice of farmers or other beneficiary groups, and involve with dominant policy narratives. This emphasizes the value of analyzing breeding institutions before planning breeding improvements, and the usefulness of path-dependency for such an analysis (Shawn, 2008).

12.3   ENVIRONMENT-FRIENDLY SORGHUM

Indiscriminate usage of herbicides to control weeds may cause environmental problems. Traditional methods of weed control are dependent on weather, costly, and labor intensive. Mature sorghum herbage has a number of water-soluble secondary chemical substances (allelochemicals). These sorghum allelochemicals can be extracted after soaking in water for 24 h and used as a natural herbicide (sorgaab). Two field trails were conducted to test the allelopathic effects of sorghum. In first experiment, sorghum stalks were incorporated in the soil and the effect of sorgaab foliar spray on the various Rabi weeds growth and on the growth and yield of wheat were considered. In the second experiment, the effects of sorgaab concentration (5 and 10%) and sprays frequency on the weeds and wheat growth were compared with hand weeding and chemical herbicides. The results revealed that the sorgaab controlled 35–49% of weeds and improved wheat yield by 10–21%. 40–50% weeds control and 15% increase in wheat yield was obtained by the incorporation of sorghum chopped herbage (2–6 Mg ha^{-1}) in the soil at sowing. Two foliar sprays of 10% sorgaab at 30 and 60 days after sowing was found to be most efficient method of weed control in wheat with highest net benefits. Therefore, the study suggested that sorgaab can be used to as a natural herbicide to control weeds in wheat (Cheema and Khaliq, 2000).

Allelopathy is sustainable and ecofriendly method of weed and disease management. An experiment was carried out to study the allelopathic effect of mulberry and sorghum water extracts (SWE) against invasive winter weeds like *Phalaris minor, Chenopodium album* L., *Avenafatua* L. and *Convolvulus arvensis* L. in wheat crop. When compared with the control, better weed control and increased yield was noticed in treatments sprayed with SWE and mulberry water extract (MWE) at the rate of 18 L ha⁻¹. The combined application of SWE and MWE had 51–55% weed control and 28%increased grain yield. Though the Atlatis (Standard herbicide) exhibited 66–68% weed control and 32% increase in final yield, it was not found economical. Since hand weeding and herbicide application were extravagant owing to their high cost and low net return, the study recommended the combined application of SWE and MWE at the rate of 18 L ha⁻¹ as best combination to control weeds and maximize the net returns in wheat (Shahbaz et al., 2018).

Host plant resistance is a feasible and cost-effective measure to reduce the effects of parasitic weed striga on field crops. In this study, a paradigm for striga resistance breeding was defined on the basis of an increased understanding of the biological basis of host-parasite interaction and by reducing the effect of environmental. The essential signals exchanged between host and parasite were identified and characterized, to determine the potential sites for intervention. The exclusive mechanisms of striga resistance associated with production of low germination stimulants (lgs), low haustorial factor (lhf), the hypersensitive response (HR), and incompatible response (IR) induction following infection were identified. Finally, by means of this method, sorghum germplasm with improved sources of resistance were identified and resistant genes were effectively introgressed into widely adapted sorghum cultivars. These new striga resistant sorghum cultivars were developed and released officially for extensive cultivation in 12 African countries (Gebisa, 2003).

Allelopathy is an eco-friendly way of controlling weeds which reduces the application of synthetic herbicides that causes environmental pollution and herbicide resistance problems. An experiment was carried out to study the effect of sorghum aqueous extracts on weeds in sunflower. The sunflower crop was sprayed with sorghum aqueous extracts individually and combined with decreased levels of Dual Gold® (S-Metolachlor) at 50, 70, and 90 DAS. The reduced density of weeds was observed in plots sprayed with sorghum aqueous extracts at rate of 15 L/ha combined with

1/3rd S-Metolachlor at 1.6 L/ha. The maximum achene yield was also realized in the plots applied with three foliar sprays of aqueous sorghum extracts together with less doses of S-Metolachlor. These results unveiled that allelopathy provides an eco-friendly and economical approach for weed management in sunflower that reduces the dependency and cost of herbicides (Sabir et al., 2016).

Phytoremediation is a low-cost method of using plants to extract, degrade, or immobilize pollutants from the contaminated environment. A successful phytoremediation depends on the selection of the ideal plant species and appropriate improving measures. In this chapter, the phytoremediation potential of sorghum was investigated. The sorghum plant was inoculated with lead-tolerant fungus (LTF) and grown in a multiple heavy metal contaminated soil with Pb, Ni, and Cu for the study. The findings revealed that the sorghum tolerated the heavy contamination. The inoculation of LTF stimulated the plant growth and increased the phytoextraction yields of Pb, Ni, and Cu. The phytoextraction potential (μg/plant) of alone sorghum was found to be 410 (Pb), 74 (Ni), and 73 (Cu). The phytoextraction potential of sorghum inoculated with LTF was 590 (Pb), 120 (Ni), and 93 (Cu) μg/plant. Therefore, the study proposed that sorghum would be the perfect plant for phytoremediation of contaminated soil due to its high phytoremediation potential, large biomass production, and application in biofuel production (Kokyo et al., 2015).

12.4   RESEARCH ADVANCES IN SORGHUM

Sorghum and pearl millet are considered as jewels of Africa. Although conventional breeding and modern marker-assisted selection made significant advances in the improvement of these species, the recombinant DNA technology may assist in broadening the gene pool by transferring genes controlling well-defined traits between species. In this chapter, the present status of sorghum and millet transformation technology was reviewed, and applications of recombinant technology in the enhancement of nutritional quality and the resistance to pathogens and pests for crops cultivated in Africa and Asia were deliberated. Further, regulatory aspects including gene flow and future potentials were discussed.

A new era in providing required tools for marker-assisted selection (MAS) has been heralded with the discovery and use of DNA markers. SSRs and modern single nucleotide polymorphism (SNP) markers, available all over the genome, have become markers of preference for gene/quantitative trait loci (QTL) mapping and for MAS/marker-assisted breeding (MAB). The speed of these activities was accelerated by the sequencing of sorghum whole genome and its availability on public domain. The candidate gene markers are being developed rapidly for different traits. The economically significant traits such as grain yield and other agronomic traits, quality traits, and resistance to abiotic and biotic stresses are potential targets of MAS. During past decades, considerable advancement has been made in the development of genomic resources in sorghum. Several QTL related to agronomic, biotic, and abiotic stress traits were identified through QTL mapping. Validation and fine mapping of QTL present a prospect to utilize MAS for sorghum improvement (Madhusudhana, 2019).

Genetic improvement of sorghum crop for a variety of economically significant traits depends mainly on understanding of genetic control of the traits. Knowledge of genetics or genetic control of sorghum traits permits sorghum breeders to create techniques and strategies for trait selection to bring out the desired genetic improvement in the target traits. Therefore, understanding the genetics of essential traits is of principal importance for quick and efficient improvement in any crop. In this chapter, the genetic control of different economic traits such as grain yield, its component traits, and yield delimiting stress factors that include biotic and abiotic factors were presented. This helps to employ effective breeding strategies to achieve greater advancement in sorghum genetic enhancement programs (Madhusudhana, 2019).

Regulation of symmetrical cell growth in the culm is essential for proper development of culm. So far, the contribution of gibberellin (GA) in this process is not confirmed in sorghum. Reynante L. Ordonio et al. (2014) mentioned that the loss-of-function mutation in four genes (*SbCPS1*, *SbKS1*, *SbKO1*, *SbKAO1*) concerned with the early steps of GA biosynthesis results in GA deficiency. This deficiency causes severe dwarfism and abnormal culm bending. Histological analysis of the bent culm unveiled that the asymmetrical cell proliferation between the lower and upper sides of culm internodes was the main cause of intrinsic bending. GA treatment reduced the bending and dwarfism in mutants, but the GA biosynthesis inhibitor, uniconazole, induced such phenotypes in wild-type plants depending on the concentration of GA. This indicated a key role of GA in regulating erectness of the sorghum culm. In conclusion, they proposed that owing to the strong association between GA deficiency-induced dwarfism and culm bending in sorghum, mutations associated with GA has not been selected in the history of sorghum breeding, as suggested from previous QTL and association studies on sorghum plant height that could not locategenes related to GA.

Sorghum crop was genetically transformed using *Agrobacterium tumefaciens*. The target explants used were immature embryos of a public (P898012) and a commercial line (PHI391) of sorghum. LBA4404 was used as *Agrobacterium* strain. This strain contains a `Super-binary' vector with *a bar* gene as a selectable marker for herbicide resistance in the plant cells. To set up a standard for conditions to be used in stable transformation experiments, a series of parameter tests was performed. Several diverse transformation conditions were assessed and from 6175 embryos total 131 stably transformed events were made in these two sorghum lines. Statistical analysis revealed a significant effect of embryos source on transformation efficiency. Transformation frequency of field-grown embryos was higher than greenhouse-grown embryos. The integration of the T-DNA into the sorghum genome was verified by the southern blot analysis of DNA from leaf tissues of T_0 plants. Herbicide resistance screening was done to confirm Mendelian segregation in the T_1 generation. Although earlier studies stated that the *Agrobacterium* method produces a higher frequency of stable transformation than other methods, the first successful application of *Agrobacterium* for production of stably transformed sorghum plants was reported in this study (Zuo-Yu et al., 2000).

Considerable amounts of genomic sequence information have been generated by crop genome sequencing projects and the availability of advanced bioinformatics tools augmented the application of this information in crop improvement programs. Here, the potential of direct application of sequence data from a sorghum genome-sequencing project in practical crop breeding programs were presented. All openly available SSR markers were aligned on a sequence-based physical map of sorghum based on sequence homology. Relating this physical map with formerly present linkage map(s) presents better opportunities for applied molecular breeding programs. Whenever new markers set is generated, the new markers can be first aligned on a sequence-based physical map and the markers located near the quantitative trait locus (QTL) can be detected from this map. This reduces the number of markers to be tested to make out polymorphic flanking markers for the QTL for a particular donor × recurrent parent combination. Polymorphic markers that are predicted (based on their location on the sequence-based physical map) to be strongly related to the target can be utilized for foreground selection in MAB. A set of markers representative of the entire genome, which would give better resolution in diversity analyses and further linkage disequilibrium mapping could be recognized with the help of this map. By targeting the precise genomic regions of interest the gaps in present linkage maps and fine mapping can be filled. Further, this opens up novel and exciting prospects for comparative mapping and for the improvement of new genomic resources in associated crops, both of which are lagging behind in the present genomic revolution (Ramu et al., 2010).

Schloss et al. (2002) collected and analyzed the DNA sequence data for 789 previously mapped RFLP probes from *sorghum bicolor* (L.) Moench Using BLAST algorithms DNA sequences, including 894 non-redundant contigs and end sequences were searched against three Gen Bank databases, nucleotide (nt), protein (nr) and EST (dbEST). Further, identical ESTs were searched against nt and nr. To find out if functional domains/motifs were congruent with the proteins known in earlier searches, translated DNA sequences were searched against the conserved domain database (CDD). In at least one of the Gen Bank searches, significant matches were observed for more than half (500/894 or 56%) of the query sequences. On the whole, proteins recognized for 148 sequences (17%) were constant among all searches, of which 66 sequences (7%) had congruent coding domains. The existence of SSRs was also evaluated in the RFLP probe sequences and 60 SSRs were developed and examined in a collection of sorghum germplasm containing in breds, landraces, and wild relatives. In comparison sorghum SSRs isolated by library hybridization screens had higher levels of polymorphism ($D = 0.69$, averaged over 38 polymorphic loci) than these SSR loci ($D = 0.46$, averaged over 51 polymorphic loci). This effect was maybe due to the comparatively small proportion of dinucleotide repeat-containing markers (42% of the total SSR loci) obtained from the DNA sequence data. Further, these dinucleotide markers had shorter repeat motifs than those isolated from genomic libraries. Based on BLAST results, 24 SSRs (40%) were detected within, or adjacent to the earlier annotated or hypothetical genes. The location of 19 of these SSRs relative to putative coding regions was determined and SSRs located in coding regions were found to be less polymorphic ($D = 0.07$, averaged over three loci) than those from gene flanking regions, UTRs, and introns

(D = 0.49, averaged over 16 loci). The sequence information and SSR loci acquired in this study will be helpful for use in sorghum genetics and improvement, in addition to gene discovery, MAS, diversity, and pedigree analyses, comparative mapping and evolutionary genetic studies.

A high degree of polymorphism in SSR markers could help in molecular dissection of agriculturally important traits in sorghum (*Sorghum bicolor* (L.) Moench). Jun-Ichi et al. (2009) designed 5599 non-redundant SSR markers, with regions flanking the SSRs, in whole-genome shotgun sequences of sorghum line ATx623. Of all SSRs 26.1% was constituted by (AT/TA)n repeats followed by (AG/TC)n at 20.5%, (AC/TG)n at 13.7% and (CG/GC)n at 11.8%. The locations identified through electronic PCR with the predicted positions of 34 008 gene loci were compared to determine the chromosomal locations of 5012 SSR markers. Most SSR markers showed similar distribution to the gene loci. In sorghum line BTx623 among 970 markers validated by fragment analysis, PCR amplification was successfully obtained for 67.8% (658 of 970) markers. In combinations of 11 sorghum lines and one Sudan grass (*Sorghum sudanense* (Piper) Stapf) line an average polymorphism rate was 45.1% (297 of 658) for all SSR loci. The product of 5012 and 0.678 indicated that SSR polymorphisms could be identified by using ~3400 SSR markers and those more than 1500 (45.1% of 3400) markers could unveil SSR polymorphisms in Sorghum lines combinations.

Pereira et al. (1994) developed a RFLP genetic linkage map of sorghum from F2 population which was obtained by crossing *Sorghum bicolor* ssp. bicolor ('CK60') and *sorghum bicolor* ssp. drummondii ('PI229828'). The map consisted of 201 loci. The loci were found to be spread among 10 linkage groups (LGs) and covered a map distance of 1530 cm. The average map distance between the adjacent loci was 8 cm. To describe the loci 55, 136, and 10, maize genomic probes (52), maize cDNA probes (124) and sorghum genomic probes (10) were used, respectively. Ninety-five percent of the loci fit predicted segregation ratios. The loci with distorted segregation ratios were restricted almost completely to a region of one LG. Sorghum and maize maps were compared and high correspondence between the two genomes in terms of loci order and genetic distance were observed. Several loci linked in maize (45 of 55) were found to be linked in sorghum. They detected cases of both conserved and rearranged locus orders.

12.4.1 REGULATION OF ETHYLENE PRODUCTION IN SORGHUM

The sorghum (*Sorghum bicolor* L. Moench) cultivar 58M contained the null mutant phytochrome B gene. It exhibited reduced photoperiodic sensitivity and a shade-avoidance phenotype. Scott et al. (1998) studied the ethylene production in the seedlings of wild type and phytochrome b mutant cultivars every three hours. These produced ethylene in a circadian rhythm. The peak production of ethylene appeared during the day. Rhythmic peaks of ethylene production of 10 times high in amplitude were produced by the phytochrome B mutant. Ethylene production in the mutant seemed to be produced from the shoot. Further, it was observed that light or the temperature cycles enabled in the production of ethylene in the mutant in diurnal rhythms, though the temperature signal appeared to override the light signal during this diurnal rhythm of ethylene production. Further, the seedlings that were grown under thermoperiods reversed with the photoperiod produced ethylene peaks during the warm nights. Their findings have established that phytochrome B is not required for proper functioning of circadian timing, but it may be taking part in modulating physiological rhythms determined by the biological clock oscillations.

High phenotypic variation has been reported in sorghum for traits affecting grain quality. The identification of genetic variants fundamental to this phenotypic variation enabled plant breeders to develop genotypes with improved grain characteristics. In this study, multiple sorghum mapping populations were phenotyped across two environments for five major grain quality traits: amylose, starch, crude protein, crude fat, and gross energy. Several QTLs that form main targets to improve grain quality for food, feed, and fuel products were revealed from coordinated association and linkage mapping. In sorghum RIL population the major QTL for crude fat content was detected. In grain sorghum diversity panel, the DGAT1 locus, a gene involved in maize lipid biosynthesis was detected. Another QTL for multiple grain quality traits like starch, crude protein, and gross energy was mapped on chromosome 1. These genetic regions identified in the study offer tremendous opportunities to improve grain composition and establish forthcoming studies for gene validation (Richard et al., 2017).

Sorghum (*Sorghum bicolor* (L.) Moench) has been a focus of plant genomics research due its significance as one of the world's important cereal crops, a biofuel crop of high and increasing importance, a progenitor of one of the world's most noxious weeds, and a botanical model for many tropical grasses with complex genomes. A rich history of genome analysis, ending in the recent with the completion of sequencing of the leading inbred genome, set a base for stimulating progress toward linking sorghum genes to their functions. The mechanisms, levels, and patterns of evolution of genome size and structure may be understood by further characterizing the genomes other than saccharine cereals. This would lay the groundwork for additional study of sugarcane and other economically important members of the group (Paterson and Andrew, 2008).

Sorghum (*Sorghum bicolor*) is a multipurpose and an ideal C4 crop for research. High-quality genome sequence has been made available for the elite inbred line BTx623. However, the limited availability of genomic and germplasm resources for inclusive analysis of induced mutations made functional validation of genes challenging. In this chapter, BTx623 seeds were mutagenized with EMS to generate 6400 pedigreed M_4 mutant pools using single-seed descent method. More than 1.8 million canonical EMS-induced mutations, influencing >95% of genes in the sorghum genome were unveiled from the whole-genome sequencing of 256 phenotyped mutant lines. Most of the induced mutations (97.5%) were different from natural variations. Further, reverse genetics was performed to determine the usefulness of the sequenced sorghum mutant resource and eight potential genes affecting drought tolerance were identified. Out of eight genes, three genes exhibited biallelic mutations and two genes had exact cosegrega-tion with the phenotype of interest. Thus, study findings established that a comprehensive resource of sequenced pedigreed mutants would accelerate sorghum breeding by providing an efficient platform for functional vali-dation of genes in sorghum. Furthermore, by using integrated genomics approaches these findings made in sorghum could be readily translated to other members of the Poaceae (Yinping et al., 2016).

To ensure food security the adaptation of staple crops should be improved in developing countries. In the West African country of Niger, sorghum is a staple crop and cultivated across different agroclimatic zones, but the genetic basis of local adaptation has not been explained. In this study, the genomic diversity of sorghum from Niger was characterized to find the genomic regions conferring local adaptation to agroclimatic zones and farmer preferences. Total 516 Nigerien sorghum accessions with known local variety name, botanical race, and geographic origin were analyzed. They found 144 299 single nucleotide polymorphisms (SNPs) by means of genotyping-by-sequencing (GBS). Discriminant analysis of principal components (DAPC) was performed and six genetic groups were identified. Afterwards loci with high discriminant loadings were identified by performing genome scan. The maximum discriminant coefficients were detected on chromosome 9, adjacent to the putative ortholog of maize flowering time adaptation gene Vgt1. Subsequently, diversity among local varieties was characterized and a genome scan of pair wise F ST Values was used to find SNPs linked with specific local varieties. Comparison of varieties termed as light and dark-grain revealed variation near Tannin1, the key gene responsible for grain tannins. These results could ease the genomics-assisted breeding of sorghum varieties that are locally adapted and preferred by farmer in Niger (Maina et al., 2018).

The sorghum [*sorghum bicolor* (L.) Moench] inbred line BTx623 has been used as a parent to develop several mapping populations. Additionally, it is used as a source to generate of DNA libraries for physical mapping, and as the inbred line selected for sorghum genome sequencing. As genetic mapping, physical mapping, and genome sequencing are all based on the same inbred line, these genetic resources have made the genome study of sorghum very effective. However, when compared with other model species, a mutant population is one important genetic resource still missing in the sorghum research community. Several avenues of research, particularly those aiming at functional genomics and bioenergy research will be opened by a systematically annotated mutant population. In this chapter, the generation of a sorghum mutant population obtained from the inbred line BTx623 by treating it with the chemical agent ethyl methanesulfonate (EMS) was reported. The mutant population had 1,600 pedigreed M_3 families; each of them was a derivative of separate M_1 seed. Several lines exhibited characteristics such as brown midrib (bmr), erect leaves (erl), multiple tillers (mtl), and late flowering (lfl), which are useful for bioenergy research. The phenotyping and genotyping results suggested that this mutant population would be an effective and useful genetic resource for both sorghum functional genomics and bioenergy research (Xin et al., 2009).

Sorghum is a major source of food and fodder for many parts of the world, particularly in arid and semi-arid regions. Its tolerance to drought/heat stresses and its strong potential as a bioenergy feedstock made it a focus of research in recent times. Further, new opportunities for the sorghum functional genomics were opened with the completion of the sorghum genome sequence. However, the accessibility of genetic resources, particularly mutant lines, is limited. Chemical mutagenesis of sorghum germplasm, followed by screening for mutants with transformed agronomically important traits, denotes a quick and effectual means of tackling this limitation. By means of Targeting Induced Local Lesion IN Genomes (TILLING) method mutations induced in new genes of interest can be effectively determined. The sorghum-inbred line BTx623 was treated with chemical agent EMS and a mutant population of 1,600 lines was generated. Several phenotypes with transformed morphological and agronomic traits were noticed in M_2 and M_3 lines in the field. A subset of 768 mutant lines was studied by TILLING with four target genes. Total five mutations were found with an estimated mutation density of 1/526 kb. Two of the mutations detected by TILLING and confirmed by sequencing were spotted in the gene encoding caffeic acid O-methyltransferase (COMT) in two separate mutant lines. The two mutant lines segregated for the predicted brown midrib (bmr) phenotype, a characteristic related with altered lignin content and increased digestibility. Therefore, the diversity of the mutant phenotypes seen in the field, and the density of induced mutations computed from TILLING indicated that this mutant population denotes a valuable resource for members of the sorghum research community. Furthermore, it has been established that TILLING is applicable for sorghum functional genomics by assessing a small subset of the EMS-induced mutant lines (Xin et al., 2008).

Semi-dwarfing genes have improved lodging resistance in sorghum which resulted in increased crop productivity. The spontaneous mutation, *dw1* observed in Memphis in 1905, was the first extensively used semi-dwarfing gene in grain sorghum breeding. In this study, the detection and characterization of *Dw1* was reported. Quantitative trait locus (QTL) analysis and cloning was performed. Results unveiled that a novel uncharacterized protein is encoded by a *Dw1*. Semi-dwarfism exhibited by rice and Arabidopsis due to the knockdown or T-DNA insertion lines of orthologous genes was comparable to that of a near isogenic line (NIL) with *dw1* (NIL-*dw1*) of sorghum. A histological analysis of the NIL-*dw1* showed that the longitudinal parenchymal cell lengths of the internode were almost similar between NIL-*dw1* and wild type. However, there was significant reduction in the number of cells per internode in NIL-*dw1*. NIL-*dw1dw3*, with both *dw1* and *dw3* that are involved in transportation of in auxin had a synergistic phenotype. These findings confirmed that the *dw1* reduces the cell proliferation activity in the internodes, and the synergistic effect of *dw1* and *dw3* improves lodging resistance and mechanical harvesting in sorghum (Miki et al., 2016).

Development of transgenic plants with altered seed storage protein composition and improved nutritive value is one of the major focuses of genetic engineering research. This task is vital for sorghum also which is an exceptional drought-tolerant cereal crop with comparatively poor nutritive value than other cereals. The resistance of one of its seed storage proteins, γ-kafirin, present in the outer layer of endosperm protein bodies, to protease digestion is considered to be one of the causes for low nutritive value of sorghum grain. Transgenic sorghum plants with a genetic construct for RNAi silencing of the γ-kafirin gene were developed using *Agrobacterium*-mediated genetic transformation. The plants with nearly floury or altered endosperm texture of kernels were obtained in T_1 generation. Reduced vitreous endosperm layer covered with thin layer of floury endosperm was found in these kernels. When compared with original non-transgenic line, 2.9–3.2 times reduction in the amount of undigested protein was observed in transgenic plants from the T_3 generation. The digestibility index reached 85–88%, in comparison with 59% in the original line. In T_2 families, the plants with high *in vitro* protein digestibility (IVPD) and vitreous endosperm type were found. There was significant reduction in the electrophoretic spectra of endosperm proteins of transgenic plants with improved digestibility, the proportion of 20 kD protein that is encoded by the γ-kafirin gene. HPLC analysis unveiled the reduction in total amino acid content and 1.6–1.7 time's increment in lysine content in two out of the three studied transgenic plants from the T_2 generation. The study results demonstrated that the development of sorghum lines with high protein digestibility and vitreous endosperm that has a high breeding value is feasible (Elkonin et al., 2016).

An EMS-induced recessive lesion *dropdead1-1* (*ded1*), that mimic mutant of sorghum was described. It is characterized by the development of spreading necrotic lesions that share many characteristics with the maize *lethal leaf spot1* (*lls1*) and Arabidopsis *accelerated cell death1* (*acd1*) mutation. They showed that similar to *lls1*, *ded1* lesions initiates by wounding and require light for constant propagation and that loss of chloroplast integrity causes *ded1* cell death. Based on this they demonstrated that *ded1* is an ortholog of *lls1* and encodes pheophorbide *a* oxidase (PaO) with 93% identity at the protein level. It was found that a stop codon-inducing single base pair changes in exon 6 of the sorghum ortholog of *lls1 forms* the mutant *ded1* allele. The *ded1* transcript was quickly and transiently induced after wounding and significantly raised in leaves having *ded1* lesions. As PaO is a major enzyme of the chlorophyll degradation pathway, its dysfunction would bring about the accumulation of pheophorbide, a potent photosensitizer that produces singlet oxygen. Consistent with this, singlet oxygen was the most probable cause of cell death related with *ded1* lesions as the role of superoxide and H_2O_2 was excluded. The signal responsible for the propagation of lesions affecting both *ded1* and *lls1* lesions was explored and it was found that both developmental age and ethylene increased the rate of lesion expansion in both mutants (Anoop et al., 2018).

Sorghum is a short-day (SD) plant and its utilization in grain or biomass production in temperate regions is determined by its flowering time regulation. Therefore, far much information is not available on the basic molecular mechanism of floral transition in sorghum. In this study, sorghum flowering locus T (SbFT) genes were characterized to construct a molecular road map for understanding the mechanism. Using multiple approaches, out of 19 PEBP genes SbFT1, SbFT8 and SbFT10 were recognized as potential candidates for encoding florigens. In phylogenetic analysis SbFT1 clustered with the rice Hd3a subclade, whereas SbFT8 and SbFT10 clustered with the maize ZCN8 subclade. These three genes were expressed in the leaf at the floral transition initiation stage. The expression of genes was early in grain sorghum genotypes but late in sweet and forage sorghum genotypes. SD induction of these three genes in sensitive genotypes was completely reversed by 1 week of long-day treatment, and still some aspects of the SD treatment had a small effect on flowering in long days, this indicated a complex photoperiod response mediated by SbFT genes (Tezera et al., 2016).

In maize The *Opaque-2* (O2) gene encodes a transcriptional activator of the b-ZIP class. In this study, a gene related in sequence to the *O2* gene of maize was isolated from sorghum and characterized. Only a single copy of gene was found in sorghum. They determined the genomic and cDNA sequences of the *O2*-related sorghum gene. Both in the promoter and in the coding region the sequence was found to be highly homologous to maize *O2*. In a total of 122 residues the b-ZIP domain with only 11 amino acid substitutions was observed in most closely related sequences. In transient expression assays, the *O2*-related sorghum coding sequence, expressed from a CaMV 35S promoter, activated expression from the maize b-32 promoter as efficiently as that obtained with the maize *O2* sequence (Livia et al., 1994).

Sorghum (*Sorghum bicolor*) is a major crop and a C_4 type plant grown worldwide. Domesticated sorghum includes many forms sweet cultivars with juicy stems and grain sorghum with dry, pithy stems at maturity. The *Dry* locus, which regulates the pithy/juicy stem trait, was found over a century ago. In this chapter, *Dry* gene that encodes a plant-specific NAC transcription factor was found. In sweet sorghum deletion or loss-of-function mutation of *Dry* resulted in cell death and modified secondary cell wall composition in the stem. Among wild sorghum and its relatives twenty-three *Dry* ancestral haplotypes, all with dry, pithy stems, were identified. Two of these haplotypes were noticed in domesticated landraces and four additional *dry* haplotypes with juicy stems were found in improved lines. These findings showed that selection for *Dry* gene mutations was a foremost step that leads to the origin of sweet sorghum. The *Dry* gene is preserved in many cereals and modification of its regulatory network could make available a molecular tool to regulate crop stem texture (Li-Min et al., 2018).

Root architecture, length of the growing season, and stomatal conductance to water vapor (g_s) are major factors having a direct effect on water inputs in cropping systems. As deep-rooted cultivars can access water from deeper layers of the soil, they perform better under water-stress conditions. Under water-limited environments reductions in *gs* reduces transpiration rate (E) and conserves water throughout the vegetative phase that may be utilized during grain filling. Furthermore, cultivation of early-maturing varieties in regions that depends on soil-stored water is a basic water management strategy. To understand the genetic basis underlying root depth, growing season length and g_s quantitative trait locus (QTL) study was conducted. In sorghum a novel QTL for crown root angle was detected on chromosome 3. A QTL for g_s was located on chromosome seven. In a follow-up field study it was found that the QTL for g_s was linked with reduced E but not with net carbon assimilation rate (A) or shoot biomass. Guard-cell length or stomatal density did not show any significant difference among the lines. This lead to the conclusion that observed differences in g_s must be elucidated by partial closure of stomata. Previously reported maturity gene *Ma1* was detected in the QTL for maturity. The probable interaction of *Ma1* with other loci explained the transgressive segregation of the population. At last, the most apparent location of the genes underlying the QTLs and candidate genes was suggested (Jose et al., 2016).

In sorghum (*Sorghum bicolor*) breeding, time to maturity decides whether a variety can be cultivated in a particular cropping system or ecosystem. Information about the nucleotide variation and the mechanisms of molecular evolution of the maturity genes would be useful for breeding programs. Here, nucleotide diversity of Ma_3, an important maturity gene in sorghum was analyzed. 252 cultivated and wild sorghum genotypes were used for the study. Based on race- and usage-based group's nucleotide variation and diversity were examined. To explore a selective sweep, in 185 of these genotypes 12 genes in the region of Ma_3 gene were sequenced. Purifying selection was found to be a strongest force on Ma_3, as low nucleotide diversity and low-frequency amino acid variants were detected. However, a very special mutation, explained as $ma_3{}^R$, appeared to be under positive selection, as suggested by considerably reduced nucleotide variation both at the loci and at adjacent regions among genotypes carrying the mutations. Further, the association study on Ma_3 nucleotide variations, revealed three important SNPs for the heading date at a high-latitude environment (Beijing) and 17 at a low-latitude environment (Hainan). The study findings provided an insight into the evolutionary mechanisms of the maturity genes in sorghum that would be valuable in sorghum breeding (Yan et al., 2015).

Sorghum's diverse germplasm collection, adaptation to harsh environments and value for comparing the genomes of grass species made it an important target for plant genomic mapping. A combined genetic and physical map of the sorghum genome (750 Mbp) was constructed in this project. For this, a novel high-throughput PCR-based method was developed to build BAC contigs and to locate BAC clones on the sorghum genetic map. Around 184 pools of BAC DNA were produced by pooling 24,576 sorghum BAC clones (approximately 4x genome equivalents) in six different matrices. Using amplified fragment length polymorphism (AFLP) technology DNA fragments from each pool were amplified, set on a LI-COR dual-dye DNA sequencing system, and examined with Bionumerics software. About, each set of AFLP primers amplified 28 single-copy DNA markers that could be used to detect overlapping BAC clones. About 2400 BACs were detected and 700BAC contigs were ordered using data from 32 different AFLP primer combinations. With the same primer pairs sorghum RIL mapping population was analyzed and about 200 of the BAC contigs were detected on the sorghum genetic map. To test and extend the contigs built using this PCR-based methodology, restriction endonuclease fingerprinting of the entire collection of sorghum BAC clones was applied. 3366 contigs each comprising an average of 5 BACs were identified by analyzing fingerprint data. BACs in nearly 65% of the contigs aligned by AFLP analysis contained enough overlap to be established by DNA fingerprint analysis. Furthermore, 30% of the overlapping BACs aligned by AFLP analysis presented information for merging contigs and singletons that could not be combined using only fingerprint data. Therefore, the integration of fingerprinting and AFLP-based contig assembly and mapping presents a consistent, high-throughput method for the construction of combined genetic and physical map of the sorghum genome (Klein et al., 2000).

Sorghum bicolor is closely associated with maize and is a staple food in Africa and many parts of the world owing to its higher tolerance of arid growth conditions. Methylation filtration (MF) technology was used to create sequence from the hypomethylated part of the sorghum genome. The evidences suggested that 96% of the genes have been sequence tagged, with an average coverage of 65% through their length. Strangely, this level of gene detection was achieved after producing a raw coverage of less than 300 megabases of the 735-megabase genome. MF specially captures exons and introns, promoters, microRNAs, and SSRs, and reduces interspersed repeats, hence giving a robust outlook of the functional parts of the genome. The MF sequence set of sorghum is valuable to research on sorghum and a potent resource as well for comparative genomics among the grasses and throughout the whole plant kingdom. The sorghum dataset supported thousands of hypothetical gene predictions in rice and Arabidopsis. Genomic resemblances emphasize evolutionarily conserved regions that will bring an improved understanding of rice and Arabidopsis (Bedell et al., 2005).

Sorghum roots releases two types of biological nitrification inhibitors (BNIs)-hydrophilic-BNIs and hydrophobic-BNIs. Previous studies showed that rhizosphere pH and plasma membrane (PM) H + ATPase are functionally associated with the release of hydrophilic BNIs, but the triggering mechanisms are not fully explained. Here, three sorghum genetic stocks were utilized to understand the regulatory mechanisms of BNIs release in root systems. Sorghum plants were grown in a hydroponic system with pH of nutrient solutions varying from 3.0~9.0. Pharmacological agents [(fusicoccin and vanadate) and anion-channel blockers niflumic acid (NIF) and anthracene-9-carboxylate (A9C)] were added to root exudate solutions. Then luminescent *Nitrosomonas europaea* bioassay was used to determine BNI activity. Sorgoleone levels in root exudates and H + excretion from roots were verified. Root PM was isolated with two-phase partitioning system and H + ATPase activity was ascertained. A reduction in rhizosphere pH increased the release of hydrophilic-BNIs from roots of all the three sorghum genotypes; however, it did not show any effect on the release of hydrophobic-BNIs. Fusicoccin promoted H + extrusion and stimulated the release of hydrophilic-BNIs. Contrastingly, vanadate suppressed H + extrusion and reduced the release of hydrophilic-BNIs. Anion-channel blockers improved the H^+-extrusion and hydrophilic-BNIs release. The study has shown that some unknown membrane transporters operate the release of protonated BNIs, which may compensate for charge balance when transport of other anions is suppressed using anion-channel blockers. So, a new proposition was put forward for the release of hydrophilic-BNIs from sorghum roots (Di et al., 2018).

Association mapping is a powerful approach for detecting genes underlying quantitative traits in plants. Alexandra et al. (2008) collected panel of 377-sorghum accessions representative of all major cultivated races and important U.S. breeding lines and their progenitors to characterize genetic and phenotypic diversity. Accessions were phenotyped for eight traits and 47 SSR loci were used to assess the levels of population structure and familial relatedness. Considerable morphological variation was observed in panel and genotypic difference between the converted tropical and breeding lines was insignificant. The performance of some association models in regulating spurious associations was assessed with the help of phenotypic and genotypic data. The analysis revealed that association models that contributed for both population structure and kinship performed better than those that did not. Furthermore, it was found that the optimum number of subpopulations utilized to correct for population structure was dependent on trait. Although expansion of the genotypic data with additional SSR loci may be essential, the association models, genotypic data, and germplasm panel explained here presented an initial point for sorghum researchers to start association studies of traits and markers or candidate genes of interest.

Stay green is a key drought resistance trait for sorghum production. QTLs for this trait with constant effects across diverse environments would increase the efficiency of selection due to its comparatively low heritability. Genome mapping and stay green evaluation was conducted on 160 recombinant inbred (RI) lines obtained by crossing QL39 and QL41. Phenotypic data were collected in replicated field trials from five locations and in three growing seasons. Data was analyzed by fitting suitable models to account for spatial variability and to explain the genotype by environment interaction. Three regions associated with stay green each in a separate LG were identified through interval mapping and non-parametric mapping in more than one trial, and two regions in single trial. The regions on LGs B and I were found to be consistent in all three trails. The multiple environment testing assisted in precise identification of QTLs associated with stay green (Tao et al., 2000).

Kong and Dong (2000) sequenced 51 clones isolated from a size-fractionated genomic DNA library of *Sorghum bicolor* (L.) Moench, that had been probed with four radio-labeled di- and tri-nucleotide oligomers. One or more simple-sequence repeats (SSRs) [72% of which were (AG/TC) n SSRs] were present in 50 clones. To amplify 38 unique SSR loci, primer sets were developed through polymerase-chain-reaction. In 18 sorghum accessions and the parents of a RI mapping population 38 loci were genotyped. The genotyping results revealed polymorphism at 36 loci in 18 accessions and at 31 loci (not including null alleles at two loci) between the parents of the RI population. Later all the 31 loci were mapped. To determine δ^* T, the estimated level of allelic difference (the estimated possibility that two members of a population, selected at random and without replacement, vary in allelic composition), at each of the loci the genotypes at 17-mapped SSR loci were assayed in 190 *S. bicolor* accessions. The average δ^* T value for *S. bicolor* was 0.89. For ten *S. bicolor* races mean δ^* T values ranged from 0.88 to 0.83 and for ten working groups (= sub-races) of the race caudatum, with only two exceptions, it ranged from 0.87 to 0.79. Among the ten race-caudatum working groups, the lowest δ^* T values for six of the loci ranged from 0.86 to 0.70. Therefore, the likelihood that diverse alleles will be present at one or more of these loci in two accessions selected at random from a working group is >0.996 when three of the loci are genotyped, and >0.9999 when all six of the loci are genotyped. The findings established that most *S. bicolor* SSR loci are adequately polymorphic and can be applied in MAS programs. Further, they indicated that the levels of polymorphism at some loci are high enough to permit the vast majority of *S. bicolor* accessions, even accessions within working groups, can be differentiated from one another by determining the genotypes at a small number, maybe as few as a half-dozen, SSR loci.

Dinakar (2000) reported the development, testing, and application (for genetic mapping) of a large number of polymerase chain reaction (PCR) primer sets that amplify DNA SSR loci of *sorghum bicolor* (L.) Moench Most of the primer sets were developed from clones isolated from two sorghum bacterial artificial chromosome (BAC) libraries and three enriched sorghum genomic-DNA (gDNA) libraries. Some were developed from sorghum DNA sequences present in public databases. Radio-labeled di- and trinucleotide oligomers were used as probes for the libraries. The BAC libraries were probed with four and six oligomers, respectively, and the enriched gDNA libraries with four and three oligomers, respectively. Both types of libraries were significantly enriched for SSRs comparative to a size-fractionated gDNA library studied previously. However, 13% and 17% of the sequenced clones acquired from the BAC and enriched gDNA libraries, respectively, lacked a SSR, whereas only 2% of the sequenced clones acquired from the size-fractionated gDNA library had no SSR. Primer sets were made for 313 SSR loci. In a population containing 18 diverse sorghum lines, 266 (85%) of the loci were amplified and 165 (53%) of the loci showed polymorphism. (AG/TC)n and (AC/TG)n repeats contained 91% of the dinucleotide SSRs and 52% of all of the SSRs a Primer sequences were described for the 165 polymorphic loci and for 8 monomorphic loci that contain a high degree of homology to genes. 66% of the trinucleotide SSRs at the loci was present in four types of repeats. The genetic map positions of 113 new SSR loci (including four SSR-containing gene loci) and a linkage map with 147 SSR loci and 323 RFLP (restriction fragment length polymorphism) loci were also reported. In one LG the number of SSR loci ranged from 8 to 30. The distribution of SSR loci was even all over approximately 75% of the 1406-cM linkage map. Though, in segments of five LGs covering about 25% of the map SSR were either few or absent. Therefore, mapping of SSR loci isolated from BAC clones positioned to these segments is expected to be the most efficient method for placing SSR loci in the segments.

Sorghum bicolor is a genetic model for C_4 grasses because of its comparatively small genome (about 800 Mbps), diploid genetics, diverse germplasm, and colinearity with other C_4 grass genomes. Here, data from deep sequencing, genetic linkage analysis, and transcriptome was used to create and annotate a high-quality reference genome sequence. To improve reference genome sequence an additional sequence of 29.6 Mbps was incorporated, the number of annotated genes increased from 24% to 34% and error frequency was decreased 10-fold to 1 per 100 kbps order. Subtelomeric repeats with characteristics of tandem repeats in miniature (TRIM) elements were found at the ends of most chromosomes. Nucleosome occupancy estimates detected nucleosomes located proximately downstream of transcription initiation sites and at various densities across chromosomes. Around 7.4 M single nucleotide polymorphisms (SNPs) and 1.9 M indels were detected by aligning more than 50 resequenced genomes from different sorghum genotypes to the reference genome. In euchromatin significant different features were recognized with periodicities of about 25 kbps. During the juvenile, vegetative, and reproductive phases, 47 RNA-seq profiles were collected from growing and developed tissues of the major plant organs (roots, leaves, stems, panicles, and seed) and a transcriptome atlas of gene expression was constructed. Analysis of the transcriptome data specified that the transcriptional profile clustering was affected largely by tissue type and protein kinase expression. The updated assembly, annotation, and transcriptome data denote a source for C_4 grass study and crop improvement (McCormick et al., 2018).

The type II CRISPR/Cas system from Streptococcus pyogenes and its simplified spin-off, the Cas9/single guide RNA (sgRNA) system, have become apparent as powerful novel tools for targeted gene knockout in bacteria, yeast, fruit fly, zebrafish, and human cells. In this chapter, adaptations of these systems resulting in effective expression of the Cas9/sgRNA system in two dicot plant species, Arabidopsis, and tobacco, and two monocot crop species, rice, and sorghum was described. In Arabidopsis and tobacco genes encoding Cas9, sgRNA, and a non-functional, mutant green fluorescence protein (GFP) were integrated using *Agrobacterium tumefaciens*. A CAS9/sgRNA complex targeted 5' coding regions in mutant GFP gene and cleaved it successfully. At the same time, error-prone DNA repair formed functional GFP genes. At the target siteCas9/sgRNA-mediated mutagenesis was confirmed by DNA sequencing. Successful expression of the Cas9/ sgRNA system in model plant and crop species shows potential for its application as a simplistic and powerful method of plant genetic engineering for scientific and agricultural purposes in future research programs (Jiang et al., 2013).

The CRISPR/Cas9 is a potent genome-editing tool which has a great prospective of improving agronomically important traits. In cereals, several studies of CRISPR/ Cas9 have been reported predominately on rice and few on other cereals like maize, wheat, and barley. Conversely, there are only a couple of reports on sorghum so far. Here, biolistic bombardment method was utilized to study the CRISPR/Cas9 system for sorghum genome editing. Using CRISPR/Cas9 two target genes, cinnamyl alcohol dehydrogenase (CAD) and phytoene desaturase (PDS), have been studied. In the sorghum genotype Tx430 effective genome editing has been attained. Further, the successful editing of the target gene with CRISPR/Cas9 was confirmed by sequencing the PCR product of transgenic plants. Both homozygosis and heterozygosis editings of CAD gene have been reported in T_0 transgenic lines PCR products. The investigation of T_1 generation of CRISPR plants unveiled that the edited gene has transferred to next generation. Three factors: (1) an efficient transformation system, (2) the design of targeted gene sequence for gRNA, (3) effective expression of CRISPR components that include Cas9 and gRNA were found as crucial elements in establishing an efficient CRISPR/Cas9 system for genome editing in sorghum (Guoquan et al., 2019).

12.5 CONCLUSIONS

The conventional breeding methods have been advanced with the application of modern breeding techniques and biotechnology tools to acquire plants with desired traits such as improved yield, disease resistance, and better quality. The modern biotechnology tools like genetic engineering and genome editing enormously increased the precision and reduced the time with which desired variations in plant features can be achieved. In this chapter, the current advancements in breeding and biotechnology methods to improve sorghum are summarized.

KEYWORDS

- **biotechnology**
- **breeding**
- **participatory plant breeding**
- **research advances**
- **simple sequence repeats**
- **sorghum**

REFERENCES

Adeline, B., Monique, D., Eric, G., Doyle, M. K., & Hélène, I. J., (2007). Local genetic diversity of sorghum in a village in northern Cameroon: Structure and dynamics of landraces. *Theoretical and Applied Genetic, 114*, 237–248.

Agrama, H. A., & Tuinst, M. R., (2003). Phylogenetic diversity and relationships among sorghum accessions using SSRs and RAPDs. *African Journal of Biotechnology, 4*, 12–19.

Alexandra, M. C., Gael, P., Patrick, J. B., Sharon, E. M., William, L. R., Mitchell, R. T., Cleve, D. F., & Stephen, K., (2008). Community resources and strategies for association mapping in sorghum. *Crop Science, 48*, 25–32.

Andekelile, M., & Zacharia, M., (2018). Evaluation of yield performance of sorghum (*Sorghum bicolor* L. Moench) varieties in Central Tanzania. *International Journal of Agronomy and Agricultural Research (IJAAR), 13*, 8–14.

Anne, G. T., Richard, M., Julius, O., & Mercy, K., (2014). The role of varietal attributes on adoption of improved seed varieties: The case of sorghum in Kenya. *Agriculture and Food Security, 3*, 9–17.

Anoop, S., Janick-Buckner, D., Brent, B., John, G., Usha, Z., Brian, P. D., & Gurmukh, S. J., (2018). Propagation of cell death in *dropdead1*, a sorghum ortholog of the maize *lls1* mutant. *Plos One, 10*, 25–34.

Aruna, C., & Deepika. C., (2019). Genetic improvement of grain sorghum. *Breeding Sorghum for Diverse End Uses* (pp. 157–173). Woodhead Publishing Series in Food Science, Technology, and Nutrition.

Bedell, J. A., Budiman, M. A., Nunberg, Andrew, C., Robert, W. R., Dan, J., Joshua, F., et al., (2005). Sorghum genome sequencing by methylation filtration. *PLoS Biology, 9*, 324–331.

Brhane, G., (1981). Salient features of the sorghum breeding strategies used in Ethiopia. *Ethiopian Journal of Agricultural Sciences, 5*, 25–34.

Cheema, Z. A., & Khaliq, A., (2000). Use of sorghum allelopathic properties to control weeds in irrigated wheat in a semi-arid region of Punjab. *Agriculture, Ecosystems and Environment, 79*, 105–112.

Di, T., Tingjun, A., Muhammad, R. Y., Tadashi, D., Santosh, Z., Yiyong, S., & Guntur, V., (2018). Further insights into underlying mechanisms for the release of biological nitrification inhibitors from sorghum roots. *Plant and Soil, 89,* 312–320.

Dinakar, B., Jianmin, D., Ashok, K. C., & Gary, E. H., (2000). An integrated SSR and RFLP linkage map of *Sorghum bicolor* (L.) Moench. *Genome, 43,* 988–1002.

Eberhart, S. A., & Russell, W. A., (1966). Stability parameters for comparing varieties. *Crop Science, 6.*

Elkonin, L. A., Italianskaya, J. V., Domanina, I. V., Selivanov, N. Y., Rakitin, A. L., & Ravin, N. V., (2016). Transgenic sorghum with improved digestibility of storage proteins obtained by *Agrobacterium*-mediated transformation. *Russian Journal of Plant Physiology, 63,* 678–689.

Fuad, A., Samuel, T., Zeleqe, L., Fikadu, T., Alemayehu, B., & Taye, T., (2018). Evaluation of early maturing sorghum (*Sorghum bicolor* (L.) Moench) varieties, for yield and yield components in the lowlands of eastern Hararghe. *Asian Journal of Plant Science and Research, 8,* 40–43.

Gebisa, E., (2003). *Integrating Biotechnology, Breeding, and Agronomy in the Control of the Parasitic Weed Striga spp. in sorghum.* Department of Agronomy, 915 West State Street, Purdue University, West Lafayette, IN 47907–2054.

Ghebru, B., Schmidt, R., & Bennetzen, J., (2002). Genetic diversity of Eritrean sorghum landraces assessed with simple sequence repeat. *Theoretical and Applied Genetics, 105,* 229–236.

Guoquan, L., Jieqing, L., & Godwin, I. D., (2019). Genome editing by CRISPR/Cas9 in sorghum through biolistic bombardment: Methods and protocols. *Methods in Molecular Biology (Clifton, N. J.), 1931,* 169–183.

Hailekiros, T., & Tileye, F., (2013). Analysis of genetic diversity of *Sorghum bicolor* ssp. Bicolor (L.) Moench using ISSR Markers. *Asian Journal of Plant Sciences, 12,* 61–70.

Iyanar, K., Babu, C., Kumaravadivel, N., Kalamani, A., Velayudham, K., & Sathia, B. K., (2015). A high yielding multicut fodder Sorghum CO 31. *Electronic Journal of Plant Breeding, 6,* 54–57.

Jiang, Wenzhi, Z., Huanbin, B., Honghao, F., Michael, Y., Bing, W., & Donald, P., (2013). Demonstration of CRISPR/Cas9/sgRNA-mediated targeted gene modification in arabidopsis, tobacco, sorghum, and rice. *Nucleic Acids Research, 4,* 78–84.

Jose, R. L., John, E. E., Patricio, M., Ana, S., Terry, J. F., & Wilfred, V., (2016). QTLs associated with crown root angle, stomatal conductance, and maturity in sorghum. *The Plant Genome, 5,* 385–391.

Jun-ichi, Y., Tsuyu, A., Tatsumi, M., Shigemitsu, K., Takashi, M., & Masahiro, Y., (2009). Development of genome-wide simple sequence repeat markers using whole-genome shotgun sequences of sorghum (*Sorghum bicolor* (L.) Moench). *DNA Research, 16,* 187–193.

Klein, P. E., Klein, R. R., Cartinhour, S. W., Ulanch, P. E., Dong, J. O., Jacque, A. M., Daryl, T. S., et al., (2000). A high-throughput AFLP-based method for constructing integrated genetic and physical maps: Progress toward a sorghum genome map. *Genome Research, 7,* 89–96.

Kokyo, O., Tiehua, C., Hongyan, C., Xuanhe, L., Xuefeng, H., Lijun, Y., Shinichi, Y., & Sachiko, T., (2015). Phytoremediation potential of sorghum as a biofuel crop and the enhancement effects with microbe inoculation in heavy metal contaminated soil. *Journal of Biosciences and Medicines, 3*, 9–14.

Kong, L., Dong, J., & Hart, G. E., (2000). Characteristics, linkage-map positions, and allelic differentiation of *Sorghum bicolor* (L.) Moench DNA simple-sequence repeats (SSRs). *Theoretical and Applied Genetics, 101*, 438–448.

Kumar, Are, A. R., Belum, V. S., Hari, C. H., Charles, T. R., Pinnamaneni, S., Ramaiah, Bhavanasi, R., & Pulluru, S., (2011). Recent advances in sorghum genetic enhancement research at ICRISAT. *American Journal of Plant Sciences, 2,* 589–600.

Li-Min, Z., Chuan-Yuan, L., Hong, L., Xiao-Yuan, W., Zhi-Quan, L., Yu-Miao, Z., Hong, Z., et al., (2018). Sweet sorghum originated through selection of *Dry*, a plant-specific NAC transcription factor gene. *The Plant Cell, 9*, 78–84.

Livia, P., Simona, L., Hans, H., Nadia, L., Vincenzo, R., Rama, J., Richard, D. T., et al., (1994). Structural and functional analysis of an *Opaque*-2-related gene from sorghum. *Plant Molecular Biology, 24*, 515–523.

Madhusudhana, R., (2019). Genetics of important economic traits in sorghum. *Breeding Sorghum for Diverse End Uses* (pp. 141–156). Wood head Publishing Series in Food Science, Technology, and Nutrition.

Madhusudhana, R., (2019). Marker-assisted breeding in sorghum. *Breeding Sorghum for Diverse End Uses* (pp. 93–114). Wood head Publishing Series in Food Science, Technology, and Nutrition.

Maina, F., Fanna, B., Sophie, M., Sandeep, R. H., Zhenbin, W., Jianan, M., Aissata, A., et al., (2018). Population genomics of sorghum (*Sorghum bicolor*) across diverse agroclimatic zones of niger. *Genome, 5*, 65–78.

Mc, C., Ryan, F. T., Sandra, K. S., Avinash, J., Jerry, S., Shengqiang, S., David, K., & Megan, A., (2018). The *Sorghum bicolor* reference genome: Improved assembly, gene annotations, a transcriptome atlas, and signatures of genome organization. *Plant Journal, 4*, 212–219.

Miki, Y., Haruka, F., Ko, H., Araki-Nakamura, S., Ohmae-Shinohara, K., Akihiro, F., Masako, T., et al., (2016). Sorghum *Dw1*, an agronomically important gene for lodging resistance, encodes a novel protein involved in cell proliferation. *Scientific Reports, 6,* 28366.

Paterson, A. H., (2008). Genomics of sorghum. *International Journal of Plant Genomics, 5,* 67–74.

Pereira, M. G., Lee, M., Bramel, C. P., Woodman, W., Doebley, J., & Whitkus, R., (1994). Construction of an RFLP map in sorghum and comparative mapping in maize. *Genome, 37*, 36–43.

Ramu, P., Deshpande, S. P., Senthilvel, S., Jayashree, B., Billot, C., Deu, M., Ananda, L., & Hash, C. T., (2010). *In silico* mapping of important genes and markers available in the public domain for efficient sorghum breeding. *Molecular Breeding, 26*, 409–418.

Reynante, L. O., Yusuke, I., & Asako, H., (2014). Gibberellin deficiency pleiotropically induces culm bending in sorghum: An insight into sorghum semi-dwarf breeding. *Scientific Reports, 4,* 5287.

Richard, E. B., Brian, K. P., Elizabeth, A. C., Bradley, L. R., Kelsey, J. Z., Matthew, T. M., Zachary, B., et al., (2017). Genetic dissection of sorghum grain quality traits using diverse and segregating populations. *Theoretical and Applied Genetics, 130*, 697–716.

Sabir, H. S., Ejaz, A. K., Hussain, S., Nadeem, A., Jabbar, K., & Ghazanfar, U. S., (2016). Allelopathic sorghum water extract helps to improve yield of sunflower (*Helianthus annus* L.). *Pakistan Journal of Botany, 48*, 1197–1202.

Schloss, S., Mitchell, S., White, G., Kukatla, R., Bowers, J., Paterson, A. S., & Kreso, S., (2002). Characterization of RFLP probe sequences for gene discovery and SSR development in *sorghum bicolor* (L.) Moench. *Theoretical and Applied Genetics, 105*, 912–920.

Scott, A. F., In-Jung, L., & Page, W. M., (1998). Phytochrome B and the regulation of circadian ethylene production in sorghum. *Plant Physiology, 116,* 17–25.

Shahbaz, K., Sohail, I., Faisal, M., & Muhammad, N., (2018). Combined application of sorghum and mulberry water extracts is effective and economical way for weed management in wheat. *Asian Journal of Agriculture and Biology, 6*, 221–227.

Shawn, J. M., (2008). Path-dependency in plant breeding: Challenges facing participatory reforms in the Ethiopian sorghum improvement program. *Agricultural Systems, 96,* 139–149.

Tao, Y. Z., Henzell, R. G., Jordan, D. R., Butle, A. D. G., & Kelly, McIntyre, M. L., (2000). Identification of genomic regions associated with stay green in sorghum by testing RILs in multiple environments. *Theoretical and Applied Genetic, 100*, 1225–1232.

Tezera, W. W., Fei, Z., Lifang, N., Shweta, K., Bhatnagar-Mathur, P., Michael, G. M., & Million, T., (2016). Three flowering locus *T*-like genes function as potential florigens and mediate photoperiod response in sorghum. *New Phytologist, 210*, 946–959.

Vinutha, K. S., Rayaprolu, Yadagiri, K., Umakanth, A. V., Patil, J. V., & Srinivasa, R. P., (2014). Sweet sorghum research and development in India: Status and prospects. *Sugar Technology, 9*, 78–83.

Xin, Zhanguo, L. W., Ming, B., Noelle, A. B., Gloria, F., Cleve, P., Gary, B., & John, (2008). Applying genotyping (TILLING) and phenotyping analyses to elucidate gene function in a chemically induced sorghum mutant population. *BMC Plant Biology, 9*, 78–86.

Xin, Zhanguo, W., Ming, L. B., Gloria, B., & John, B., (2009). An induced sorghum mutant population suitable for bioenergy research. *Bioenergy Research, 2,* 10–16.

Yan, W., Lubin, T., Yongcai, F., Zuofeng, Z., Fengxia, L., Chuanqing, S., & Hongwei, C., (2015). Molecular evolution of the sorghum maturity gene *Ma$_3$*. *Plos One, 14,* 51–62.

Yinping, J., John, B., Ratan, C., Gloria, B., Junping, C., Bo, W., Chad, H., et al., (2016). A sorghum mutant resource as an efficient platform for gene discovery in grasses. *The Plant Cell, 28*, 1551–1562.

Zuo-yu, Z., Tishu, C., Laura, T., Mike, M., Ning, W., Hong, P., Marjorie, R., et al., (2000). Success of sorghum breeding-agrobacterium-mediated sorghum transformation. *Plant Molecular Biology, 44*, 789–798.

Index